NATO AND TERRORISM

On Scene: New Challenges for First Responders and Civil Protection

T0140487

NATO Science for Peace and Security Series

This Series presents the results of scientific meetings supported under the NATO Programme: Science for Peace and Security (SPS).

The NATO SPS Programme supports meetings in the following Key Priority areas: (1) Defence Against Terrorism; (2) Countering other Threats to Security and (3) NATO, Partner and Mediterranean Dialogue Country Priorities. The types of meeting supported are generally "Advanced Study Institutes" and "Advanced Research Workshops". The NATO SPS Series collects together the results of these meetings. The meetings are co-organized by scientists from NATO countries and scientists from NATO's "Partner" or "Mediterranean Dialogue" countries. The observations and recommendations made at the meetings, as well as the contents of the volumes in the Series, reflect those of parti-cipants and contributors only; they should not necessarily be regarded as reflecting NATO views or policy.

Advanced Study Institutes (ASI) are high-level tutorial courses intended to convey the latest developments in a subject to an advanced-level audience

Advanced Research Workshops (ARW) are expert meetings where an intense but informal exchange of views at the frontiers of a subject aims at identifying directions for future action

Following a transformation of the programme in 2006 the Series has been re-named and re-organised. Recent volumes on topics not related to security, which result from meetings supported under the programme earlier, may be found in the NATO Science Series.

The Series is published by IOS Press, Amsterdam, and Springer, Dordrecht, in conjunction with the NATO Public Diplomacy Division.

Sub-Series

A.	Chemistry and Biology	Springer
B.	Physics and Biophysics	Springer
C.	Environmental Security	Springer
D.	Information and Communication Security	IOS Press
E.	Human and Societal Dynamics	IOS Press

http://www.nato.int/science
http://www.springer.com
http://www.iospress.nl

Series B: Physics and Biophysics

NATO AND TERRORISM
On Scene: New Challenges for First Responders and Civil Protection

edited by

Frances L. Edwards
San José State University
Department of Political Science,
San José, CA, U.S.A.

and

Friedrich Steinhäusler
University of Salzburg
Division of Physics and Biophysics,
Salzburg, Austria

Published in cooperation with NATO Public Diplomacy Division

Proceedings of the NATO Advanced Research Workshop on
Emergency Management after a Major Terror Attack: The New Challenges for
First Responders and Civil Protection,
Ericeira Portugal
2–4 March 2006

A C.I.P. Catalogue record for this book is available from the Library of Congress.

ISBN 978-1-4020-6276-6
ISBN 978-1-4020-6277-3 (eBook)

Published by Springer,
P.O. Box 17, 3300 AA Dordrecht, The Netherlands.

www.springer.com

Printed on acid-free paper

All Rights Reserved
© 2007 Springer
No part of this work may be reproduced, stored in a retrieval system, or transmitted
in any form or by any means, electronic, mechanical, photocopying, microfilming,
recording or otherwise, without written permission from the Publisher, with the exception
of any material supplied specifically for the purpose of being entered
and executed on a computer system, for exclusive use by the purchaser of the work.

Table of Contents

Foreword

The present security environment is and will be complex and global, and subject to unforeseeable developments. We realize that international security developments have an increased impact on the lives of the citizens of our home countries.

Terrorism, increasingly global in scope and lethal in results, and the possible spread of weapons of mass destruction are likely to be the principal threats to civil populations over the next 10 to 15 years.

Peace, security and stability are more interconnected than ever. This places a premium on close cooperation and coordination among relevant international organizations, playing their respective, interconnected roles.

The NATO Alliance will continue to follow the broad approach to security of the 1999 Strategic Concept and perform the fundamental security tasks it set out, namely security, consultation, deterrence and defense, crisis management, and partnership.

Collective defense will remain the core purpose of the Alliance. The character of potential Article 5 challenges, the article of the North Atlantic Treaty on collective defense, is continuing to evolve. Large scale aggression against the Alliance will continue to be highly unlikely; however, as shown by the terrorist attacks on the United States in 2001, following which NATO invoked Article 5 for the first time, attacks may originate from terrorist groups and involve unconventional forms of armed assault. Future attacks could also involve the use of weapons of mass destruction. Therefore, defense against terrorism and the ability to contribute to the protection of civilian populations, critical infrastructure, and to support consequence management will retain an increased importance for NATO.

NATO and Partners' populations and critical infrastructures may face various threats, including the use of Chemical, Biological, Radiological and Nuclear (CBRN) weapons. It is a requirement of the Alliance to help enhance the ability to deter, disrupt, defend and protect against terrorism, and more particularly to contribute to the protection of the Alliance's population, territory, critical infrastructure and forces and to support consequence management.

We all know and we all agree that response to a disaster or a catastrophic event is first and foremost a national responsibility. Stricken nations that cannot cope with the situation within own resources can request international assistance. Arrangements are in place within the framework of international organizations to respond to such a request. Some of these arrangements are based on solidarity others are more of a binding nature with international agreements.

Increased cooperation with the United Nations and the European Union will continue to be a priority for the Alliance. In this respect, the Riga Summit Declaration, in line with the Istanbul Summit Communiqué, states that NATO-EU relations cover a wide range of issues of common interest, including civil emergency planning, and calls for "improvements in the NATO-EU strategic partnership as agreed by our two organizations to achieve closer cooperation and greater efficiency and avoid unnecessary duplication, in a spirit of transparency and respecting the autonomy of the organizations".

The first workshop in 2004 took a close look at the threats first responders face when dealing with the consequences of terrorism against civilian populations. This second workshop continued this important work by examining existing Emergency Management Structures. While the emphasis of the second workshop clearly was on national structures as they are the ones that have the responsibility to respond to catastrophic events, the workshop also provided an opportunity to look at international arrangements that are available to complement and supplement national response.

Guenter W. Bretschneider
Head of the Euro-Atlantic Disaster Response
Coordination Centre at
NATO Headquarters Brussels

Preface

On March 2–4, 2006 NATO held a Joint STS-CNAD Workshop, entitled *Emergency Management After a Major Terror Attack: The New Challenges for First Responders and Civil Protection,* held in Ericeira, Portugal. The meeting was organized jointly by Prof. Friedrich Steinhäusler (University of Salzburg, Austria), and Dr. Frances Edwards (San Jose State University, California, U.S.A.). Altogether 25 experts from 11 countries participated, presenting an overview of the current situation with regard to first responder and emergency management capabilities within the different countries.

The contributions represent the opinion of the authors but not necessarily that of the institution they represented at the NATO meeting.

F. Steinhäusler, F. Edwards

Acknowledgements

The logistical support by the following institutions is gratefully acknowledged: NATO (Public Diplomacy Division), Brussels (Belgium); University of Salzburg, Salzburg (Austria); San Jose State University, California (U.S.A.); Bioterrorism Unit of Interpol, Lyon, France; California State University, Northridge, California (U.S.A.); Norman Y. Mineta Transportation Institute, San Jose, California; NewRisk, Limited, London, England; Norwegian Institute of International Affairs, Oslo, Norway; Grand Valley State University, Grand Rapids, Michigan; Respond BV, Tilburg, The Netherlands; EMERCOM, Moscow, Russia; State Fire Department and Emergency Medical Service of Hamburg, Germany; United States Environmental Protection Agency, Boston, Massachusetts; and Menlo Park Fire Protection District, Menlo Part, California.

The sponsorship provided by NATO Scientific Committee is also gratefully acknowledged.

Furthermore, the editors wish to acknowledge the valued assistance provided by Ms. Lyudmila Zaitseva, without whom the workshop would not have been possible; and Ms. Claudia Heissl for her dedicated work in supporting the logistics of the workshop.

Contributors

BACIU, Adrian, Bioterrorism Unit, Public Safety and Terrorism Sub-directorate, Interpol, General Secretariat, 200 quai Charles de Gaulle 69 006 Lyon, France

BALLARD, James David, Department of Sociology, California State University, Northridge, 18111 Nordhoff St., Northridge, CA 91330-8318

BRETSCHNEIDER, Günter, Euro-Atlantic Disaster Response Coordination Center (EADRCC), NATO, Boulevard Leopold III, B-1110 Brussels, Belgium

GOODRICH, Daniel C., Norman Y Mineta Transportation Institute, 20405 Via Volante Cupertino, CA 95014

LEIVESLEY, Sally, NewRisk Limited, Frazer House, 32/38 Leman St. London E1 8EW, United Kingdom

MAERLI, Morton Bremer, Norwegian Institute of International Affairs, P.O. Box 8159 Dep. N-0033 Oslo, Norway

MULLENDORE, Kristine, School of Criminal Justice, Grand Valley State University, 271 C DeVos Center, 401 W. Fulton St., Grand Rapids, MI 49504-6431

OTTEN, Jan, Respond BV, Jules Verneweg 121, 5015 BK Tilburg, The Netherlands

POTAPOV, Boris, EMERCOM of Russia, Moscow, Russia

RECHENBACH, Peer, State Fire Department and Emergency Medical Service of Hamburg, Westphalensweg 1, D-20099 Hamburg, Germany

RYDELL, Stan, U.S. Environmental Protection Agency, 1 Congress St., Boston, MA 02114-2023

SCHAPELHOUMAN, Harold, Menlo Park Fire Protection District, California Task Force 3 – National Urban Search and Rescue Program, 300 Middlefield Rd, Menlo Park, CA 94025

1. INTRODUCTION

1.1. Overview

NATO member states are facing a new form of terrorism—its actors operate globally, they reject compromise and negotiations, and they are threatening to deploy weapons of mass killing. The terror attacks in New York, Washington, DC, Madrid, and London have shown clearly that the population in these countries is at risk. In view of this new risk environment, emergency planning needs to adapt to these security challenges.

International terrorism remains a significant threat to the peace of the world. Terrorists have already attacked cities and their critical infrastructure with conventional explosives and delivery systems, such as car bombs, truck bombs or suicide bombers, chemical and biological weapons. Other modes of attack are postulated to be deployed within the near future. Some of the materials to be used are unusual or exotic, while some are modifications of materials used daily for other purposes: chemical, biological, radiological, nuclear, and explosive (CBRNE). These materials pose a safety and security challenge for first responders, slowing the response and potentially killing scarce trained personnel.

This book provides information about how leading agencies across the NATO membership are preparing to confront and respond to terrorists deploying weapons of mass destruction (WMD), weapons of mass killing (WMK), and weapons of mass disruption (WMDi).

Based on contributions by experts from NATO member states, Israel, and the Russian Federation, this book offers information on fourth-generation warfare as a new challenge to first responders. It also provides a variety of insights into methods for preparing for, and responding to, WMD, WMK, and WMDi events, using new technologies and strategies. Hence, the book offers solutions for restoring the community to functioning after such major terror attacks. Topics include preparing, planning, operations, equipment, training, and unique applications of existing technologies for countering terrorist attacks against cities and their infrastructure. Short- and long-term actions are recommended in order to assist in managing the aftermath of a major terror attack. Conclusions are derived from the presentations and findings of a meeting organized with the framework of the NATO Joint Security through Science Conference of National Armaments Directors Workshop "Emergency Management After a Major Terror Attack: The New Challenges for First Responders and Civil Protection," Ericeira, Portugal, March 2–4, 2006. For the list of participants and contributors see Appendix 1.

1

F. L. Edwards and F. Steinhäusler (eds.), NATO and Terrorism, 1–6.
© 2007 *Springer.*

1.2. The Issue

Organized first responders find it difficult to respond effectively against people determined to do something symbolic, such as a suicide bombing. Terrorist attacks using relatively simple weapons and tactics demonstrate that a sophisticated society cannot protect itself against primitive weapons. Fourth-generation urban warfare relies on speed, simplicity, and a lack of concern for getting caught, while the typical first responder organization relies on complex structures which may impede speed, and an unwillingness to trade the lives of first responders for the lives of terrorists. Only the military still adheres to the doctrine of "acceptable losses" in the twenty-first century.

More robust intelligence has been the most successful defense against terrorists. Specialists assert that only information warfare keeps ahead of terrorists. First responders are about reactions, and they cannot prepare for all scenarios at the same quality level. Even if they are prepared, they cannot prevent people from dying. Information is the most crucial point for dealing with terrorist threats. Western intelligence is quite efficient, but in order to remain effective it has to maintain its covert character. With all that intelligence has prevented, there is still little public understanding of the threats that exist. The intelligence community cannot tell the public what was prevented because it would reveal their techniques and sources. The terrorists have comparatively few successes, but that is all the public sees and interprets as the intelligence community's failures. Terrorists succeed with one bomb.

In the period 2001–2005, five major terror attacks occurred on the territory of the NATO member states: Spain, Turkey, the UK, and the USA. Altogether these attacks have resulted in 3,283 civilian deaths and 4,387 injured—among them about 425 US first responders. Direct damages amounted to about €54 billion and indirect economic costs to society totaled over €188 billion. All of these terror attacks have deployed either conventional explosives, or aircraft as guided missiles. Based on international intelligence information, it can no longer be excluded that future terror attacks may also use WMDi, WMK, or WMD.

Catastrophic urban disaster events pose a challenge for first responders based upon the type, scope, and complexity of the event. Search and rescue activities are among the most important phases in the overall emergency response. In the case of an act of explosive-based terrorism this may involve structural collapse. Initial actions by first responders are critical, and proper understanding of the value, and importance, of methodical search techniques requires an often disciplined and focused response, not always found during disasters.

1.3. Emergency Planning and Terrorism

Contrasted to the fragile bipolar standoff of the cold war, the contemporary security picture is increasingly blurred and characterized by new dimensions of uncertainty. The current unpredictability coexists with an unprecedented level of vulnerability—as well as perceptions of vulnerability—as a product of modernity and modern lifestyles and demands. The emergence of spectacular and widely publicized low-probability/high-consequence events has fueled feelings of susceptibility among the population, and, because of this, a renewed interest in protection against terrorism through emergency management structures.

Some 60% of people in 35 countries believe that the war in Iraq has increased the likelihood of terrorist attacks worldwide, a recent poll reveals (BBC, 2006). Just 12% think terrorist attacks have become less likely. Publication of the cartoons of Mohammed in Denmark on September 30, 2005, and then in Norway in January 2006 stirred controversy and anger. People were killed during riots, embassies and companies of the two countries were attacked, and Danish commodities were embargoed.

Deliberate hostile acts against key targets could be staged by calculating opponents with the power to impact or impair vital societal functions and national interests. Asymmetrical threats may have become more than a military and poetical buzzword. The failure of just a few key mainstays and functions within the modern society could result in wide-scale problems. Privatization and short-term economic interests predominate, and priority is given to maximizing profit and improving efficiency, often at the expense of the necessary security and preparedness measures. The possible nexus of international terrorism and nonconventional means could create a particularly deadly mix.

In parallel, the emergency management structures put in place also need to meet and mitigate threats of more daily character high-probability/low-consequence events. This dual use of emergency management products and capabilities challenges leaders to find the budget to maintain the required level of preparedness. As Hurricane Katrina amply demonstrated in the USA, no one really knows how to define adequate levels of preparedness, even for "routine" or repeated events, like hurricanes.[1] What is the appropriate social response to allocating scarce resources and setting priorities for emergency preparedness? While civilian emergency planners have been concerned with the reallocation of all hazards program support to what they see as an exaggerated focus in defense circles on high-consequence/

---

[1] For a review of Hurricane Katrina, see Appendix 5.

low-probability events, military actors have had the opposite problem, with civilians being preoccupied with low consequences of a high probability (Dalgaard-Nielsen and Hamilton, 2006). Proper emergency management planning hence calls for an understanding of the possible spectrum of events, and their relative likelihood and overall implications. Being prepared for different scenarios can reduce the extent of the damage and increase the probability of a successful crisis management.

Terrorism often has as its objective to impose a political, economic, or religious change of regime. Through their acts, terrorists seek to destroy the fundamental interests and structures of society, and to generate fear and insecurity amongst the population, e.g. through assaults and attacks on civilians. In other words, these are crimes directed against the security and democratic values of the state and the population in general. The effort put into fighting terrorism therefore has a high priority on the national as well as the international scene. For example, in accordance with Section 114 of the Danish Penal Code, terrorism is defined as a series of very serious offenses (e.g. murder, arson, kidnapping, hijacking) committed by groups, organizations, and persons, with the purpose of frightening the population, destabilizing the order of society, or attempting to force Danish or foreign authorities and international organizations to carry out or omit carrying out a specific action. The American Federal Bureau of Investigation defines terrorism as "the unlawful use of force or violence against persons or property to intimidate or coerce a government, the civilian population, or any segment thereof, in furtherance of political or social objectives." Traditionally, there has been a distinction between internal or national terrorism, state-sponsored terrorism, and international (transnational) terrorism.

The objectives of emergency preparedness are to reduce the vulnerability of infrastructure and important industries, to minimize damage caused by crisis or war, to safeguard the life, health, and welfare of the population, and to attempt to meet the needs of the civil population and the military forces in the supply of important goods and services in crisis or war. Contingency planning for a crisis should provide for the ready disposition of available resources in a flexible manner. The goal should be to increase everyday security and safety, reduce the danger of major accidents, minimize vulnerability to the loss of vital services, improve the capacity and competence in crisis management, and preserve national security. Preventive and reparative measures may be introduced to strengthen security, through the development of a greater knowledge of risks, threats, and vulnerability; the inclusion of contingency planning and security concerns into general planning; preparations to ensure continuity of government, crisis management, and consequence management; and regular evaluation of all safety and contingency measures.

1.4. The Framework

The NATO Joint STS–CNAD Workshop on "Emergency Management after a Major Terror Attack: The New Challenges for First Responders and Civil Protection" had two purposes:

To assess the level of civil protection created by the current disaster-management infrastructure in NATO member states; and

To identify possible improvements that could be achieved through reasonable investments in training and organizational reforms.

Workshop participants have been selected based upon their expertise in the subject areas of threat assessment, emergency management, research and development in field response to terrorism, and international disaster response. Experts from NATO member states including the USA, as well as the Russian Federation and Israel, have provided altogether 23 contributions to the above topic areas. These were grouped as follows:

- Review of the international CBRNE challenge to emergency management, including the evolution to fourth-generation warfare and the legal ramifications of combatants without battlefields or uniforms

- Assessment of the current needs of first responders for improved protection, detection and search and rescue technology, and emerging technologies with emergency planning and management applications

- Assessment of the current emergency management status in selected countries, including current methods of evaluation of capacity and charting improvement plans

- Assessment of international terrorism prevention and response efforts within the EU and NATO, and the importance of developing common command and control structures, mutual aid plans, and joint training and exercising

- Plenary discussion of the emergency management areas in need of additional scientific research, technological development, or strengthening

Technical presentations on the application of Geographical Information System (GIS), air modeling, and remote sensing demonstrated the benefits of applying existing technologies to emergency planning and response to enhance capabilities and save lives. Furthermore, the US Environmental Protection Agency demonstrated the use of computer-simulation modeling as a tool to assess the consequences of a terror attack employing a chemical or radiological dispersal device (Dirty Bomb).

Papers from several of the participants are included in this book, as well as information from discussions and other contributions. Contributing authors include Adrian Baciu, James David Ballard, Günter Bretschneider, Daniel C. Goodrich, Sally Leivesley, Morton Bremer Maerli, Kristine Mullendore, Jan Otten, Boris Potapov, Peer Rechenback, Stan Rydell, and Harold Schapelhouman.

2. SECURITY THREAT ASSESSMENT

2.1. Preparing for Natural Hazards versus Terror Attack: A Matter of Risk

Currently all NATO member states, in fact the global security community, are involved in the discussion of various security threat scenarios potentially resulting in damages comparable to acts of "catastrophic terrorism," e.g. threats resulting from the deployment of a weapon with nuclear, biological, or chemical mass destruction (WMD) capability by terrorists. Whilst it is undoubtedly a worthwhile aim of many countries to prepare for the aftermath of such an event, it is difficult for any society to find the correct level of emergency preparedness. The much-needed high level of emergency preparedness is extremely costly and difficult to maintain over extended periods of time for reasons of financial constraints and political unwillingness to invest in countermeasures for an event of unknown probability. These investments have to compete with the generally acknowledged need for investing in strengthening the emergency preparedness of society in case of large natural disasters, such as floods, hurricanes, earthquakes, or tsunamis.

Many aspects of emergency preparedness lend themselves to the "all hazards" approach, in which planning, training, and exercising benefit the first responders' capabilities across all catastrophic disasters. The development of a uniform emergency response system, such as the Incident Command System; creation of an Emergency Operations Center; and development of individual residents' and community emergency response capabilities at the neighborhood level will result in lives saved in any disaster scenario.

Ideally, the level of emergency preparedness should be based on a realistic, scientifically derived *risk assessment* for the threat due to terrorism *versus* threats due to major natural disasters. Table 1 shows the number of major natural catastrophes and terror attacks, as well as the number of victims, for the period after the terror attacks in the USA on September 11, 2001 till December 2005.[1] In this period 1,360 persons were killed due to major terror attacks, compared to several hundred thousands killed by major natural disasters. Therefore, the risk of dying in the course of a major natural disaster exceeds by at least two orders of magnitude the lethal risk due to terrorism.

---

[1]Excluding terror attacks in Iraq and Israel.

F. L. Edwards and F. Steinhäusler (eds.), NATO and Terrorism, 7–48.
© 2007 *Springer.*

TABLE 1. Comparison of the number of victims due to major natural catastrophes and major acts of terrorism over the period October 2001 to December 2005

Event	Location	Year	Number of dead
6 terror attacks	Tunisia, Yemen, Bali, Russia, Kenya	2002	396
7 terror attacks	Saudi Arabia, Morocco, Afghanistan, Russia, Indonesia, Turkey	2003	188
4 terror attacks	Russia, Spain, Indonesia	2004	572
4 terror attacks	Egypt, Jordan, UK	2005	204
Earthquake	Bam, Iran	2003	26,271
Earthquake/Tsunami	Southeast Asia	2004	229,866
Floods	India	2005	1,000
Hurricane Katrina	USA	2005	1,306
Earthquake	Pakistan	2005	20,000–30,000 (estimated)

Society in NATO member states expects increasingly to be protected against *all* risks, be they natural or man-made. It has to be emphasized though that every society has only finite resources to address risk reduction from these threats. Therefore, the risk of experiencing a major terror event can only be minimized, but never reduced to zero. The main objectives of emergency planners are minimizing: (a) the number of victims; (b) damage to the infrastructure; and (c) impact on the environment. However, there is only limited experience of national governments in regaining control in the aftermath of large-scale catastrophes; below are examples of the reaction by authorities to major man-made and natural catastrophes (Steinhäusler, 2006):

- *Japan*: Response by the authorities after the US nuclear weapon attack on Hiroshima on August 6, 1945 at 8:16 a.m. was significantly delayed. Already shortly after 8:16 a.m., the control operator of the Japan Broadcasting Corporation and the Tokyo railway signal center found that there was no communication with Hiroshima. However, it took until just before 10 a.m. for the Central Command Headquarters in Osaka to report that military communications to the city had failed. By 1 p.m. the government in Tokyo suspected that Hiroshima was a dead city, but the precise cause remained unknown. Not until August 7 at 6:30 p.m. did a plane with a government delegation land at the Kichijima airfield, starting its investigation on August 8, 1945, 58 hours after the event (Wyden, 1984).

- *India*: Response by the authorities to the intentional chemical release accident in Bhopal in 1984 was plagued by the prolonged impact over extended periods. The initial response phase had to address the consequences of the release of methyl isocyanate, which killed immediately as many as 3,800 people. However, national response agencies were unprepared to deal with pronounced health effects, still being detected many years after the release (Dhara and Dhara, 2002). This experience shows the difficulty of managing an airborne chemical release, and estimating the long-term effects.

- *Brazil*: Extensive experience has been gained in the aftermath of the radiological accident contaminating the Brazilian city of Goiania in 1987, when unsuspecting thieves stole a strong radioactive medical source and subsequently contaminated citizens and the urban environment. The incident showed the multiple practical difficulties confronting the professionals and the authorities caring for the radiation victims, addressing the needs of those suspected to be contaminated, and the reaction by the uncontaminated residents of Goiania. (For details, see Appendix 3.) These experiences can be used for assessing the adequacy of emergency management after an act of radiological terrorism.

- *Former Soviet Union*: Response by the former Soviet Union to the explosion at the nuclear power plant in Chernobyl (Ukraine) in 1986, the largest nuclear accident in history, was plagued by its initial emphasis on covering up the true magnitude of the accident. Eventually recovery work involved an estimated 250,000 workers from all over the Union. The number of residents living in a *strict control zone* reached 360,000 in 1989 (Steinhäusler, 2005).

- *USA*: Despite the earlier experience of a terror attack on the World Trade Center in February 1993, the unprecedented scale of the terror attacks in New York City on September 11, 2001, overwhelmed the organizations responsible for managing the emergency response (Kean and Hamilton, 2004). Lacking situational awareness, experiencing inadequate communication, and having significant deficits in evacuation procedures added to the overall confusion of the emergency managers (details are summarized in Appendix 4).

- *USA*: Response by dedicated US organizations to the impact of Hurricane Katrina showed several deficiencies at the local, state, and federal level (for a detailed review, see Appendix 5). On August 26, 2005, Governor Kathleen Blanco declared a state of emergency in Louisiana (Governor's Office, 2005). Louisiana requested federal troops to replace their National Guard troops in Iraq (Honore, 2005), and the Federal Emergency Management Agency (FEMA) was given full authority to respond to

Katrina.[2] Nevertheless, the Mayor's decision to operate from a hotel room during the disaster, without Emergency Operations Center trained, resulted in confusion over organizational structures, with no coordination between city and federal resources as late as September 1, 2005 (CBC News, 2005).

Generally, *prior to a major catastrophe*, the willingness to invest in national emergency preparedness is low. Neither the political establishment nor the average citizen can reasonably estimate the magnitude of the consequences of such a large-scale catastrophe, be it due to a catastrophic terror attack, a technical failure in a large installation, or a natural cause. Therefore, many countries are inadequately prepared to deal with a major catastrophe. *After a major catastrophe* it is likely that society will overreact, hoping to prevent another such event with an abundance of hardware, software, and operational procedures aimed at strengthening the level of national emergency preparedness. Instead of these extremes, it is proposed here that the level of emergency preparedness should be based on a realistic, scientifically based *risk assessment*. This applies to all types of catastrophes, including major terror attacks.

The assessment of the risk (R) for a major terror attack requires the following input data:

$$R = P_1 * P_2 * C/E \qquad (1)$$

Where:
R = risk to society due to terrorists carrying out a major terror attack;
P_1 = probability that a terrorist is motivated sufficiently to carry out such a major attack;
P_2 = probability that a terrorist will succeed in carrying out a major attack;
C = consequences to man and the environment due the major terror attack;
E = effectiveness of countermeasures as part of the emergency preparedness of society aimed at mitigating the damages resulting from a major terror attack.

Since these input parameters are associated with significant uncertainties, the resulting risk assessment is also subject to large uncertainties due to error propagation.

---

[2]"Specifically, FEMA is authorized to identify, mobilize, and provide at its discretion, equipment and resources necessary to alleviate the impacts of the emergency [White House]."

The *motivation* by terrorists to carry out a major terror attack has already been demonstrated for biological and chemical WMD.[3] Therefore, it is safe to assume that the value of P_1 is approaching 1 for all three types of WMD, i.e. nuclear, biological, and chemical, at least for organizations with a motivational drive similar to that of the two terror groups Aum Shinrikyo and al-Qaeda.

Terrorists have already demonstrated that they have the *capability* to use materials suitable for biological and chemical WMD, so the probability P_2 (biological or chemical) for terrorists to have the capability to build and deploy a biological or chemical WMD also approaches 1.

With regard to nuclear capability, it is necessary to consider several stages of the logistical requirements for terrorists to fulfill. Physical security at military sites is presumed to be sufficiently high to exclude the diversion of a *military* nuclear weapon from such a storage site. Furthermore, there are several built-in safeguard features terrorist would have to overcome in order to activate a weapon. Also, the terrorists would have to know the operational status of the weapon, such as whether it has been serviced. In summary, the probability P_2 (nuclear) for terrorists to deploy an advanced nuclear weapon can be considered as extremely low.[4] In view of the many difficulties involved in acquiring or building a *crude* nuclear device, the probability P_2 (nuclear) in this case is also rather low, as long as they are denied possession of weapon-usable nuclear material. However, this assessment no longer holds, if terrorists should be able to acquire sufficient amounts of such material, particularly if they should obtain adequate amounts of highly enriched uranium (HEU). In this case P_2 (nuclear) can be assumed to be high, since it is within the realm of terrorists to acquire the scientific know-how and set up the operational logistics for building a crude, HEU-based nuclear device.

The *consequences* of a major terror attack on man and the environment differ significantly, depending on the mode of attack and the weapon deployed, with the largest impact resulting from the deployment of a WMD.

A nuclear terror attack on a NATO member state, both in terms of short- and mid-term effects on the targeted nation, would be severe, mitigated only to some limited extent by international mutual assistance programs. Irrespective of modifying factors, such as meteorological conditions at the

---

[3] The religious sect *Aum Shinrikyo* (Japan) and the *Al-Qaeda* international terror network have shown to be sufficiently motivated for such acts of terror: the former has already demonstrated its motivation in implementing two sarin attacks in Japan; the latter has announced its intentions in 2004 (for details, see Steinhäusler, 2006).

[4] It cannot be excluded that terrorists will continue to try to obtain such an advanced nuclear weapon.

time of detonation, below-ground or near-ground detonation, and population density in the target area, C (nuclear) can be assumed to be high: blast,
heat, and radiation would inflict significant harm to people and necessitate
major cleanup operations.

The consequences of a biological attack on a NATO member state
depend on the type of biological agent (e.g. infectious, length of incubation
period) and the size of the population attacked. If the attack is limited to one
region, there is a reasonable probability that the effects of the outbreak of a
disease could be curtailed to some extent; therefore, C (biological) is low.
If, on the other hand, several attacks are synchronized and occur simultaneously in different regions, this could approach pandemic dimensions,
making C (biological) high.

The consequences of a chemical attack on NATO member states are
limited as compared to C (nuclear or biological). Although the number of
victims is expected to be high, the impact would be limited to a more or less
defined area.[5] Under this assumption the consequences to the population
and the environment can be considered to be within the realm of dedicated
emergency response organizations, particularly if they were supported by
specialized units from the national armed forces and international support
units.

The *effectiveness of countermeasures* in NATO member states is subject
to considerable variation between nations, and will be subject of more detailed discussion in the following chapters. However, it can be generalized that
at present nuclear, as well as biological and chemical, WMD attacks find
member states vulnerable to a varying degree, showing partially significant
deficits in protecting even their first responders as the first line of defense
(Steinhäusler and Edwards, 2005). If the experience gained from the Bhopal
act of sabotage, the Hiroshima atomic bomb attacks, the cleanup operations
of the Chernobyl accident at the nuclear power plant, and the response
to the disaster due to Hurricane Katrina can be used to extrapolate the
effectiveness of countermeasures in case of a chemical (E [chemical]) or
nuclear (E [nuclear]) WMD attack, this would indicate that—despite the
unquestionable severity of the incident—adequately strengthened emergency
preparedness can limit the worst consequences at least to some extent, none-
theless most likely with a considerable time delay.

There is no adequate experience on the effectiveness of countermeasures in case of a biological WMD attack, except for possibly the influenza
pandemic in 1918, with several tens of millions dead. Based on the experi-

---

[5]This, of course, depends significantly on the meteorological conditions at the time of the
chemical release and the population density of the target area.

ence gained during the very limited anthrax attacks in the USA in October 2001, and the isolated cases of avian flu in Asia, Africa, and Europe, it is safe to assume that the effectiveness of emergency preparedness E (biological) for such a global incident is still low.

The fight against terrorism and its potential to deploy WMD can only be won if the efforts to prevent such acts of terrorism are balanced with the ability of society to respond to these threats in a cost-efficient manner. Inadequate emergency preparedness will result in obvious failure to protect the domestic civilian population and the national infrastructure. Excessively high expenditures for a low-probability event will not be able to be sustained over an extended period of time due to effects of "political fatigue."

Using the results of the assessment of probabilities P_1, P_2, consequences C, and the effectiveness of emergency countermeasures E in equation 1, Table 2 summarizes the results of the qualitative risk assessment for the potential deployment of nuclear, biological, or chemical WMD by terrorists. In view of the significant lack of numerical data, it is not feasible to carry out a quantitative risk assessment.

TABLE 2. Risk to society due to a WMD attack by terrorists

Type of WMD attack	P_1	P_2	C	R	Risk	Assumption
Crude nuclear device	1	High	High	Medium	High	Access to an adequate amount of weapon-grade HEU
Advanced military nuclear weapon	1	Close to zero	High	Low	Low	Adequate physical protection of weapon storage
Biological weapon	1	1	High	Low	High	Coordinated multiple releases of biological agent
Chemical weapon	1	1	Medium	Medium	Medium	Impact limited to specific area due to stable meteorological conditions

The risk (R) for society to experience a terror attack with a crude nuclear device or a biological weapon is comparatively higher than that from the deployment of an advanced military nuclear weapon; the risk from the use of a chemical WMD by terrorists is in between the other three risks.

2.2. The New Risk Environment

Emergency managers are facing a new risk environment due to new security threats: (a) Terrorists capable of inflicting casualties even at well-guarded sites due to the unconventional use of conventional explosives; (b) terrorists interested in causing mass panic with radiological weapons; (c) terrorists actively searching for ways to obtain and deploy WMDs.

2.2.1. UNCONVENTIONAL USE OF CONVENTIONAL EXPLOSIVES

Unconventional use of conventional explosives has become a major threat to the individual citizen, assets of the national critical infrastructure, and members of the civilian and military security community alike. The novel and continuously changing techniques developed by terrorists cover a wide range and are summarized below.

- **Suicide terrorism:** Suicide attacks, when people themselves are weapons of war, have risen from an average of about 3 per year in the 1980s to more than 40 a year in 2002, and almost 50 a year in 2005. By 2005 about 400 suicide bombings have taken place in Iraq alone since the invasion in March 2003. Globally 75% of all suicide bombings have happened since the terror attacks in the USA on September 11, 2001. Altogether in 16 countries over 300 suicide attacks have resulted in more than 5,300 people killed and over 10,000 wounded. Since 2003, suicide terror attacks have also been registered in Egypt, India, Russia, and the UK. Although they represent just 3% of all terrorist incidents in the period 1980–2003, they have caused 48% of all fatalities.

  In the typical modus operandi the terrorist wears an explosive belt on the body or carries it in a suitcase, sports bag, or rucksack (about 3–15 kg of TNT or homemade explosives, together with small chunks of iron or a large quantity of nails packed around explosives). The detonator of a simple design allows the terrorist to activate the explosives even when under pressure. Alternatively, a suicide terrorist drives a truck, carrying an explosive load of several tons, toward a prearranged target. Increasingly suicide terrorists work as teams, with at least one of them waiting to detonate the explosives upon arrival of the first responders on the scene of the initial terror attack.

- **Improvised explosive device (IED):** An IED is a conventional bomb, where typically several hundred grams of explosive are hidden in an unlikely container. The container can be virtually anything suitable for covertly placing the explosive, with actual examples ranging from a melon, pressure cooker, metal tube, animal carcass, hollowed-out rock, wine bottle, backpack, or human corpse. The IED can be equipped with an infrared sensor, arming the device when someone approaches the IED. Typically, the explosion is triggered by remote control (e.g. through a mobile telephone or a toy plane or car controller). Depending on the amount of explosive material hidden inside the IED and the characteristics of the target area, the number of victims can range from a few individuals to about a dozen. Emergency management is increasingly faced with the added challenge—besides having to stabilize an already disastrous scene—that multiple IEDs have been hidden by terrorists and detonated upon arrival of first responders at the scene.

- **Vehicle-borne improvised explosive device (VBIED):** A VBIED based on a car typically contains several dozen kilograms or several hundred pounds of explosives (mostly ANFO[6]). Truck bombs carry a larger load of explosives, notably vehicles with up to 8 t of explosives on board have already been detonated. The vehicle is parked near an area with increased population density (marketplace, shopping center, office building) or driven at high speed against the target (e.g. checkpoint) and detonated either by remote control or by a suicide driver on board. The type of vehicle used determines largely the number of casualties resulting from the detonation (typically several dozen). However, the Oklahoma City bombing in 1995 caused the collapse of the Murrah Federal Building and the deaths of 168 adults and children (CNN, 1995), while the bombing of the Marine Barracks in Beirut caused the deaths of 241 marines (CNN, 2003).

  A variant of this mode of attack is the use of two vehicles in a convoy: the first vehicle, usually the smaller of the two, is driven by a suicide bomber and detonated at a barrier or checkpoint protecting the targeted site, thereby creating access by destroying any barrier and causing casualties among the guards; the second vehicle, generally a truck, breaks through this opening and is able to deliver a large amount of explosives directly at the target area inside the breached protective perimeter. Alternatively, the bomb can be delivered using a motorcycle, bicycle, animal-drawn cart, boat, aircraft, or hang glider.

- **Vapor cloud or gas-air explosion (air fuel bomb):** An alternative use of vehicles is a suicide attack with a gasoline/propane tank truck, releasing up to 20 t of gasoline (gasoline/diesel 30:70 or gasoline 100%) or large amounts of propane gas. The truck is driven into the targeted site (e.g. shopping mall) and the tank is ruptured by a small amount of

---

[6]ammonium nitrate, diesel oil plus a small booster charge

explosive. Alternatively, the truck is parked at the openings of a ventilation shaft or fresh air intake, and the content of the tank is released
into the ventilation system. In case of strong intervention by police
forces the attackers destroy the tank with explosives, causing a large
explosion and fire. If the terrorists' intervention is successfully delayed,
the targeted site is flooded with gasoline or propane gas that will lead to
a massive vapor cloud or gas-air detonation and fire destroying the site.

2.2.2. RADIOLOGICAL WEAPONS

Radiological weapons can be used by terrorists in several ways, all aimed
at causing deliberate radiation exposure, thereby adding an additional layer
of complexity to emergency management of already facing a difficult
situation. The deployment of a radiological dispersal device (RDD) is likely
to result in relatively low *individual* radiation exposure of the targeted
population, in most cases insufficient to cause a severe radiation detriment.
Nevertheless, the social and psychological effects can be severe, particularly in an urban area where a large number of persons may either be
actually contaminated or perceive to be contaminated.

- **Intentional covert irradiation:** A strong radiation source is placed
  inside an inconspicuous container (e.g. a gamma radiation emitter diverted from a medical facility is hidden inside a fire extinguisher in a
  subway station). Subsequently, this device is placed in an area with high
  population density, thereby irradiating a large number of persons over a
  period of several days. Each of the individuals is likely to receive only
  a minor radiation upon passing by the hidden source. However, the
  collective dose by the large number of persons irradiated is probably
  significant. After a certain period terrorists inform the media, revealing
  the covert radiation exposure of a large population group. Upon realization by the public of potentially having been contaminated, this will
  probably cause widespread panic. Emergency management will have to
  deal with tens of thousands of people in need of assessing their actual
  degree of radioactive contamination.

- **Radiological dispersal device (RDD):** An RDD disperses radioactive
  material into the environment. This can be achieved by generating
  radioactive aerosol over an extended period of time (e.g. introducing
  covertly the aerosol into a ventilation system), or by combining the radioactive material with conventional explosives (e.g. truck bomb loaded with
  radioactive material blanketed with conventional explosives; Dirty Bomb).
  Alternatively, a fully fueled plane, loaded with radioactive material and
  conventional explosives, could be crashed intentionally into a target.

Besides the distribution of the radioactive material by the blast due to the plane crash, the heat from the fire would cause hot air, mixed with radioactive particles, to rise several hundred meters. This would increase the area and the number of persons affected by radioactive contamination. Emergency management will face radioactively contaminated victims, contaminated first responders and their contaminated equipment, as well as elevated levels of environmental radioactivity in the area surrounding the site of attack.

2.2.3. WEAPONS OF MASS DESTRUCTION

WMDs such as nuclear, biological, or chemical weapons in the hands of terrorists represent the most devastating threat scenario. So far, WMD attacks have involved the use of anthrax (several American states and cities were attacked in October 2001; the attacker had not been identified by 2007), as well as sarin (the Japanese Aum Shinrikyo cult carried out such attacks in Matsumoto in June 1994 and in Tokyo in March 1995). Emergency response units and management showed severe deficits in both cases, and lacked analytical capabilities or demonstrated inadequate incident management capabilities and communication policies. No nuclear WMD attack has occurred hitherto.

- **Improvised nuclear device (IND):** The use of an IND by terrorists is likely to occur in the following manner: (a) An IND is smuggled into a city harbor on board a container ship.[7] An ultimatum is issued that the device will be detonated unless terrorist demands are met (nuclear blackmail); (b) an IND is placed on a truck and detonated in a metropolitan area without prior warning.[8]

- **Biological WMD:** A biological WMD requires that a biological agent (e.g. anthrax) is covertly released into the environment. Methods may include (a) a biological aerosol released at night to avoid damage to the agent by ultraviolet radiation and to use a potential temperature inversion allowing the biological aerosol to remain close to the ground for extended periods; or (b) a biological agent contained in powder form in envelopes shipped to various addresses by mail. Upon opening of the envelopes persons and their environment are contaminated.

---

[7]Typically only 2–4% of containers arriving in a harbor are physically inspected for their content.

[8]The explosive yield of such a device would be a TNT equivalent of about 10,000 t at most, about 30% less than the two military nuclear devices exploded over Japan in World War II.

An infectious disease may also be released using food. The Rajneeshee Cult in The Dalles, Oregon, contaminated the salad bars there. "The commune members plotted to take over Wasco County government in 1984, spiking local salad bars in The Dalles area with salmonella in an effort to incapacitate non-Rajneeshee voters. The action sickened some 750 people and crippled the local economy as fear spread" (Parker, 2006). The former Soviet Union was perfecting "bomblets" to deliver smallpox and other infectious disease material to civilian populations during the cold war (Staten, 1999).

- **Chemical WMD:** Typically a volatile nerve agent is covertly released into the environment, representing a vapor hazard for the victims as well as the first responders. Over the last 50 years only one terror organization, Aum Shinrikyo, has deployed such a weapon successfully, although Iraq is known to have used chemical weapons against the Kurds (Raymond, 2006). Sarin dissolves in water and can be deployed in droplet sprays or dispersed in warm climates.

2.2.4. EMERGENCY MANAGEMENT CHALLENGES

Irrespective of the actual mode of WMD attack, such an attack will represent a challenge to emergency management far beyond the routine threat scenarios. The uncertainties associated with the damage assessment will be problematic: how large is the number of victims; what is the extent of physical damage; is there any concurrent level of environmental contamination; how reliable is the assessment of available resources to counter the attack? All of this will add to the already high level of stress to the emergency response management and the first responders having to address the needs of victims expected to number in the thousands.

Furthermore, in case of a WMD attack emergency management will have to provide solutions under significantly aggravated conditions compared to a conventional terror attack. These will include several unique challenges. (a) Complex search and rescue operations will have to be conducted in severely contaminated areas. Many members of emergency response units are not routinely equipped to detect radioactive, biological, or chemical WMD-related contaminants, and under routine conditions they do not wear adequate personal protective equipment (PPE). Therefore, they are likely to suffer significant losses, which in turn will impede their capabilities to find and remove victims from the attack area. (b) Speedy damage control by the emergency responders is expected as an essential objective of their action. It is questionable whether this can be achieved in the initial phase when WMD-contaminant levels may be life-threatening. (c) The area affected by the contamination is likely to change over time, depending on the weather situation in the postattack phase. This will make delineation of the area to be cordoned off, evacuated, or quarantined a challenging task.[9] (d) Securing the crime scene is considered of prime importance after the attendance to the immediate victims of the WMD attack. In view of the widespread destruction at the Ground Zero area and its impact on the C_3 capabilities[10] of the targeted community, carrying out sophisticated forensic investigations under these circumstances is expected to be difficult.

2.3. Challenges to Law Enforcement due to a Major Terror Attack

2.3.1. CHALLENGES DUE TO CBRNE THREATS

Law enforcement will play an important role in preventing, disrupting, or investigating a major terror attack. Furthermore, law enforcement agents will be among the first responders called to the scene of any major terror attack. At the time they will not know whether—in addition to the deployment of conventional explosives—hazardous nuclear (N) and other radioactive (R), biological (B), chemical (C), or explosive (E) materials will be present at the crime scene. Dispersion methods for CBRN materials may be as simple as placing a container in a heavily used area, opening a container, using conventional or commercial spray devices, or as elaborate as detonating an IED, combined with CBRN materials. Therefore, it is vital for the law enforcement agents to be able to differentiate as early as possible, if such an agent is present.

Chemical materials potentially used by terrorists include the classical warfare agents, such as nerve agents and blister agents, or toxic industrial chemicals (TICs). They may be in the form of a powder, liquid, or gas. Exposure can occur due to skin contact, ingestion, or inhalation. Some simple characteristics can assist the emergency management in deploying members of the law enforcement community to the scene of the terror attack in delineating the areas of potentially life-threatening exposure, if real-time channels of communication are provided, allowing the instant reporting of a rapid onset of symptoms within seconds to minutes upon exposure (i.e. coughing, burning skin, convulsions, death); unusual smells; presence of dead insects, birds, or fish, and/or brown withered vegetation.

Biological materials potentially used by terrorists pose a special challenge to emergency managers. The main difference between biological terrorism and conventional terrorism (such as bombs and hijackings) is the length

---

[9] See Chapter 4 for a discussion of plume modeling for scene characterization.
[10] Command–control–communication.

of time from the attack to the recognition of victims of the attack (incubation period). Depending on the agent, the incubation period for a biological attack can be up to 60 days. Because of the relatively slow onset of symptoms in the target population, the potential bioterrorist has two options in carrying out an attack: (1) the biological agent is released covertly through obvious dissemination, or inclusion of a warning or threat; or (2) the biological agent is disseminated covertly, presenting itself as a natural disease outbreak within the population. The ability to undertake the attack covertly significantly increases the probability that hospitals, not the traditional community of first responders (law enforcement, hazardous materials—HAZMAT, fire), will be the first to recognize a bioterrorism event. Most likely this will be due to the unfolding epidemiology and gradual increase in attack rates of a communicable agent, meaning that covert incidents will primarily be detected by public health officials.

In this regard, preparing for a bioterrorism response requires considering both the overt and the covert attack scenarios, and planning appropriately for each. Traditionally, police plan and exercise extensively for responding to "overt" acts of terrorism, such as bomb attacks, hostage takings, and assassinations. While acts of bioterrorism may be overt, enabling police to respond traditionally, unique challenges arise when individuals or groups use biological agents covertly. As biological agents are not detected by the senses, and the time from exposure to onset of illness may take weeks, perpetrators may be attracted to this form of crime as they have plenty of time to flee the "crime scene."

Nuclear and other radioactive materials come in a variety of physical forms, including metallic solids, powders, liquids, and gases. Exposure at the crime scene can occur through skin contact, or inhalation of incorporation (e.g. dust). The presence of ionizing radiation is not detectable by the senses because there are no visible signs of a radiation hazard at the scene of a terror attack. The situation is further aggravated for emergency management because radiation exposure may cause symptoms which are delayed for hours, weeks, or even years, depending on the mode of exposure, dose rate, and the duration of the exposure. Therefore, emergency management needs to take into consideration that a law enforcement agent present at the scene of the terror attack without adequate radiation detectors may receive potentially life-threatening doses.

A wide variety of conventional *explosive materials* are used in terror attacks, ranging from industrial-grade explosives stolen from mines to self-made ANFO,[11] and military explosives diverted by corrupt members of the

---

[11]Ammonium nitrate and fuel oil.

armed forces. The amount detonated has been increasing over the last decade: currently up to several tons have been deployed in a single terror attack hidden in a concrete mixer truck.

2.3.2. COVERT *VERSUS* OVERT TERROR ATTACK

Law enforcement agents need to account for the differing aspects of covert and overt attacks in the development of all national terrorism response protocols.

An **overt** attack is an event clearly recognizable by law enforcement or other responding agencies. There will be general awareness of *what, where, when,* and/or *how* a potential terror attack is likely to occur, using the following: (1) information on a specific threat, typically an explicit warning by the terrorists or intelligence information; (2) discovery of the means suitable for dispersal of an R, B, C, or N material, or other signature activities (e.g. specific technical or laboratory equipment; (3) discovery of questionable or suspect materials indicative of R-, B-, C-, or N-related activities. Since there is delayed onset of symptoms due to R and B material, public health officials may only become aware of any impending health consequences when they are contacted by law enforcement agents. The procedure for this "contact" should be preestablished in order to avoid delay and inefficient information exchange. The consequence of failing to have such an established information exchange protocol between law enforcement and public health for overt attacks is a significant delay for health providers to ramp up their awareness and logistical capacity for consequence management of an incident. There is also the possibility of the investigation suffering, as law enforcement will not have the benefit of knowing when victims or perpetrators seek medical treatment and their identity.

When dealing with an overt potential terrorist attack, past experience has taught that the first necessary task is to secure the area and ascertain the nature and severity of the threat. Reported examples include when a secondary device has been targeted at emergency responders,[12] or offenders have perpetrated armed secondary assault in an attempt to harm or kill rescuers and disrupt emergency operations. Therefore, in most cases, both a primary and secondary secured perimeter should be established. A thorough search of these perimeters should be a priority at the onset of the incident.

---

[12]For example, abortion clinic bombings in Georgia (USA) by Eric Rudolph. See http://www.cnn.com/SPECIALS/2005/rudolph/ for more details.

In the event of an overt CBRN release, a large up-wind area may also need to be rapidly secured for use as a staging and operating environment.

If the perpetrators are aiming for a covert attack, they will use carefully disguised dispersal means with neither an indication of an R, B, or C agent being spread, nor a threat or warning being issued. In these situations, initially victims remain unaware that they have been exposed, and law enforcement is not aware that a crime has taken place. The terrorist desires this scenario, as victims will not seek the necessary medical treatment until they experience symptoms; so for many agents this delay will mean almost certain death. In this case, a criminal investigation may only begin after public health has taken notice through a surveillance system within the health-care community or the coroner which indicates an unusual disease pattern such as one of the following:

- Sudden increase in the number of patients
- Significant number of persons with similar, usually flu-like, symptoms
- High mortality rate with victims having a common home and/or work location, or activities
- High mortality rate among the wrong patients at the wrong time of the year, e.g. a 35-year-old man dying of the flu in August
- Human victims concurrent with illness in the animal population
- Disease not normally seen in a particular geographical location or season of the year
- Occurrence of a disease with mandatory notification, such as inhalation of anthrax or smallpox.

2.3.3. RESPONSE BY LAW ENFORCEMENT AGENTS[13]

Law enforcement cannot deal with major acts of terrorism on its own. It is critical that emergency management foresees that adequate agreements are in place, and that their practical implementation is subject to regular exercises between law enforcement and partner agencies.[14] Such exercises should outline their respective and/or cooperative roles in dealing with a

---

[13]Some material regarding international law enforcement issues is based on a paper by Adrian Baciu delivered at the NATO STS-CNAD Workshop, Portugal, 2006.

[14]In the USA the Metropolitan Medical Response System (MMRS, started as the Metropolitan Medical Strike Team—MMST) was created, starting in 1997, to bring together police, fire, emergency medical services, emergency services, coroner, public health, public works and other on-scene first responders to plan for WMD/CBRNE events. For a description see http://www.chem-bio.com/resource/2001/jan2001.pdf.

major attack. Most important are agreements with the public health community that include an early warning system wherein law enforcement is informed of any emerging suspicious health issues, a means for disclosure of information to the media, accommodating patient privacy and confidentiality issues, the collection and handling of evidence, selection of compatible personal protective equipment, and connections with other national and international health organizations.

One of the present problems concerning the response to CBRNE materials is the fact that, routinely, little information sharing is taking place between differing agencies, except on an informal or individual basis. Secondly, there is no international "clearing house" or database of exercises that have been conducted, "lessons learned," outcomes of actual incidents, or model programs to emulate in planning efforts.[15] More importantly, there is little sharing of intelligence information with regard to threats of CBRNE attacks, real or imagined.

Intelligence Sharing Across Professions

Often overlooked intelligence-gathering resources are available and unused within the civilian response community. For example, in the case of bio-terrorism, intelligence can come from a number of important sources not typically considered in law enforcement, including the syndromic surveillance of disease occurrence routinely carried out by many public health organizations, and from the open-source publications within the science community. Additionally, the evaluation and analysis of raw intelligence may prevent intelligence from being shared in a timely manner. However, this analysis is critical to provide a timely perspective to the user of this intelligence. Thus, an understanding of the types of intelligence available, and the challenges of distributing useful intelligence products, must be addressed as part of the development of an information-sharing plan.

The Homeland Security Advisory Council's Intelligence and Information Sharing Initiative group issued a report in December 2004 identifying the intelligence-sharing gaps as a critical problem for the USA's counter-terrorism efforts (Homeland Security Advisory Council, 2004). The report noted that federal-level agency sharing was inadequate, and that it did not extend to the state and local agencies at all. The Lessons Learned Information Sharing staff then held four regional meetings with subject matter

---

[15]In the USA the Oklahoma City Memorial Institute for the Prevention of Terrorism has created an online resource, Lessons Learned Information Sharing, www.llis.gov. After action reports from exercises and real events are included, as well as specialized plans for pandemic flu and radiological dispersal devices, personnel from emergency services in the USA can register for passwords for access to the protected site.

experts to determine the existing needs for better intelligence sharing. More than 60 people participated from the various professions concerned with counterterrorism intelligence gathering. While identifying a series of critical gaps, the experts noted that local and regional solutions had been developed from the "bottom up" in the absence of any federal initiatives, and that they were effective, offering models for national and international application (Department of Homeland Security, 2005).

Far greater strides should be made in regard to developing viable channels of communications that would transfer applicable information to and from "the street." These collection initiatives must start with local law enforcement activity and data gathering, and include aggressive analysis and rapid dissemination to all relevant agencies, federal, state, and local. Some of the current gaps are strategic, while others are tactical. Efforts are underway to resolve both types of problems.

Currently most law enforcement agencies use "field information cards" (FI cards) to collect data on suspicious persons and activities that an officer encounters during the workday. At present, these cards are hard copy that must be transcribed into a database system for future use by other officers. While this information from "the street" is crucial when it is fresh, its value diminishes as it ages. Due to budgetary strictures in most large American cities, personnel cutbacks have occurred routinely in the last 10 years. Because of public pressure and union issues, most cities have chosen to eliminate civilian jobs rather than sworn positions when required to save money. When budget-cutting time comes for most police departments, they eliminate the civilian support positions, such as records secretaries, the people who enter the FI cards into the officer-accessible database. In one major metropolitan city, one of the USA's ten largest, the backlog of FI cards is over 6 months. Information that is being collected today will not be accessible to other officers for 6 months, at which time it will no longer be actionable.

Two solutions have been tried for the problem of the loss of street-level intelligence. The first is the development of some handheld devices that allow the officer to enter information directly into a database. This device is then placed in a cradle in the station at the end of the officer's shift, and the data are downloaded automatically into the central computer, from which other officers can access it within a few hours. However, many police officers resist the use of technology because they find it cumbersome to carry around, and the keypad is too small for them to use comfortably. Additional work is needed to develop a portable, handheld device for use, collecting information that overcomes the objections of officers who are not comfortable with the limitations of the currently available small handheld devices. Improvements in voice recognition software, for example, might eliminate the need of keyboard information.

A second solution is the development of specialized intelligence units within individual police departments, and their regional networking across departments, often through regional collection organizations sponsored by the Federal Bureau of Investigation (FBI). One of the earliest was the Bay Area Terrorism Working Group (BATWING).[16] Started in 1998 by the Counterterrorism Unit of the San Francisco Regional Office of the FBI and the California Office of Emergency Services Coastal Region Office, this organization brings together representatives from police, fire, emergency medical services, emergency services, public health, mental health, and key businesses to receive briefings from government agencies, and provide an opportunity to share information on counterterrorism initiatives. The group meets quarterly, and includes all the Metropolitan Medical Response System agencies, as well as all the mutual aid support agencies in the Bay Area.

Los Angeles created the Terrorism Early Warning Group (TEWG) in 1996. Originally this organization was more law enforcement-oriented. Its success in countering terrorist initiatives led to the creation of Operation Archangel, designed to bring the business community and the public into the intelligence-sharing process. The Los Angeles County Sheriff's Department led both groups. By 2006 the TEWG concept was being adopted across the USA using State Homeland Security Grant funding from the federal government.[17] For example, in August 2006 the State of Nevada announced the creation of two such groups, one each for the northern and the southern portions of the state: "The multi-agency groups will collect intelligence, analyze suspicious activities to detect possible patterns, and share information with local, state, and federal agencies" (Maimon, 2006).

A Joint Regional Information Exchange System (JRIES) was established by the Department of Homeland Security, providing laptop computers with intelligence collection software to the largest cities in the USA and their critical infrastructure partners. Homeland security grants also provided for the development of regional intelligence-sharing organizations based in sheriff's offices. Starting with the 2004 Law Enforcement Terrorism Prevention Program grants, funds could be used "to purchase equipment and support efficient and expeditious sharing of information and intelligence that could preempt possible terrorist attacks" (Department of Homeland Security, 2004).

---

[16]Access to the BATWING website is now password-protected. Information about law enforcement counterterrorism groups can be found at http://www.adl.org/learn/additional_resources/resources_terrorism.asp.

[17]For a comprehensive list of the regional intelligence groups nationwide see http://www.adl.org/learn/additional_resources/resources_terrorism.asp.

Opening up of cross-border information sharing within the Alliance remains a high priority for counterterrorism. Some of the multistate initiatives in the USA might provide examples of structures that could be replicated to provide for more rapid cross-border intelligence sharing.

Identifying Hazardous Radiological, Biological, and Chemical Materials

Identification of the hazardous R, B, or C material represents the second most pressing problem involving a major terror attack. Timely identification at the scene of an overt attack can be technically difficult to ascertain with any degree of certainty, especially following the release of a biological agent. However, it is of significant value to perform field screening to determine the presence of R, B, or C materials. Even if the level of identification is associated with some uncertainty at this early stage, it should be viewed as an asset in assessing the type of risk at the crime scene. The top priority in the management of a major terror attack involves the identification of physical and chemical properties of the substance released. Any information about the identity of the agent having been deployed by the terrorists will assist in selecting the appropriate emergency response procedure. Only after identifying the material can an effective plume or deposition model be developed which will define the outer perimeters of the hot and warm zones. Neutralization plans can then be formulated, decontamination procedures established, emergency medical treatment plans undertaken, and environmental damage preservation precautions foreseen. In the event of a covert attack, identification of HAZMAT is equally critical, but is likely to represent only the starting point of the response. For example, in a covert biological attack, it may be the identification of a suspicious disease cluster by the public health system that triggers an investigation.

Emergency management planners need to account for the fact that most civilian emergency service agencies, including specialized HAZMAT teams, generally do not possess effective testing equipment to help identify sophisticated biological agents which may have been deployed in a potential bioterrorist attack. Moreover, some of the existing testing equipment is not validated for accuracy in field use, where samples are often contaminated with additional substances such as soil or blood, and definitive testing of even simple agents can take time and require expertise. While they may be able to identify those agents that have civilian counterparts, there are any number of others for which they have no testing reagents or detection equipment.

In any major terror attack, the media will need to fill any information void, turning law enforcement into one of their prime initial counterparts for authoritative information. This makes it critical for law enforcement to include a media relations checklist in the advanced planning for a bioterrorism event, in order to provide reliable information to the public as quickly and accurately as possible.[18]

2.3.4. PREVENTION, DISRUPTION, AND INVESTIGATION BY LAW ENFORCEMENT AGENTS

Law enforcement agents have the prime responsibility to prevent, disrupt, and investigate terror threats and, after such an attack, bring the criminals to justice. Therefore, in addition to the development of response protocols, law enforcement should initiate a *Program of Prevention* including: (1) assisting in strengthening physical security at civilian facilities (universities, research laboratories, hospitals, etc.) which store or handle R, B, or C materials with potential use for terrorists[19]; (2) ensuring mandatory reporting of accidents, theft, loss, or release of R, B, or C materials potentially useful for terrorists; (3) promoting the restriction or monitoring of the purchase and use of dual-use laboratory equipment; (4) enforcing regulations dealing with the transport and shipment of R, B, and C materials potentially useful for terrorists; (5) increasing the level of R–B–C terrorism awareness and information exchange between the law enforcement and the scientific community.

Such a preventive program[20] should entail adequate legislation, providing investigative authority of suspected R, B, or C attacks or attempts to mount such an attack. Coordination meetings held amongst law enforcement and public health officials are useful to discuss their respective roles in dealing with an R, B, or C attack, such as addressing agreements on disclosure of information to the media, conducting joint criminal and epidemiological investigations, information sharing including patient privacy and confidentiality issues, collection and handling of evidence, selection and availability of personal protective equipment, and establishing contact points with other national and international organizations. Practice drills and/or tabletop exercises enhance further the coordination among the participating agencies and

---

[18]For a complete discussion of the role of the media in terrorism response, see Winslow, F. E., 2002, Telling it like it is: the role of the media in terrorism response and recovery. *Perspectives on Preparedness*, 9, August, http://bcsia.ksg.harvard.edu/BCSIA_content/documents/Role_of_the_Media.pdf.

[19]For a description of the US Partnership for Critical Infrastructure Security, see http://www.pcis.org/; for a description of the 2007 Infrastructure Protection Program, see http://www.ojp.usdoj.gov/odp/grants_ipp2007.htm.

[20]For example, the Interpol Bioterrorism Prevention Program, see http://www.interpol.int/Public/BioTerrorism/ and http://www.interpol.int/Public/BioTerrorism/links/default.asp.

potential responders.[21] Ideally, cooperating agencies have an early warning system in place wherein the health community notifies the law enforcement agencies of any suspicious health issues. Timely agreement on a clear command and control structure among concerned agencies is most valuable. As the MMRS model has proven, training must be provided to select dedicated CBRNE responders and officials responsible for responding/investigating potential CBRNE incidents.

Disruption of criminals planning a major terror attack, or investigating the criminal scene of such an attack, can pose significant challenges to the law enforcement agents:

- *Communication*: The terror attack may involve several national jurisdictions, even different nations, therefore requiring a high degree of coordinated communication and investigative efforts. The unusual nature of an attack involving N, R, B, or C materials will require close coordination among law enforcement, research institutions, and public health during the investigation. All communication with the media will have to be coordinated among law enforcement, research institutions, and public health in providing accurate information to the media, without compromising an ongoing criminal investigation;

- *Crowd control*: The large number of wounded and/or killed after a major terror attack involving N, R, B, or C materials will pose a significant resource constraint on normal law enforcement efforts. The breakdown of large segments of critical infrastructure after a major terror attack, together with widespread panic, will add to the difficulties of law enforcement agencies trying to maintain a minimum level of order in the affected population. Both the sick and the "worried well" are likely to quickly exceed the medical response capacity of the community, necessitating additional law enforcement capacity to maintain some degree of order at the medical response centers;

- *Resources*: Investigating law enforcement agents will need additional resources in order to acquire adequate means to minimize levels of security risks, physical risks, and health risks to themselves (Steinhäusler and Edwards, 2005). Speed of spread of attack and disease will require enormous investigative resources, with the unpredictable nature of such a scene making it difficult to preserve and analyze criminal evidence.

---

[21]In addition, facilitated exercises, such as those held by the MMRS across the USA, are helpful to test plans and enhance field response capabilities. An announcement of one such event may be found at http://www.citizencorps.gov/citizenCorps/eventDetail.do?eventID=533. For a more complete discussion of facilitated exercises, see http://www.semp.us/securitas/2006JanFeb.html.

The inherent conflict between rescuing and caring for the victims, and preserving a crime scene, must be addressed in early planning, so that agreements are already in place regarding scene control, scene access, and priorities for evidence protection and collection at, and through, medical care facilities.

2.3.5. GENERAL INDICATORS OF CBRNE TERRORIST ACTIVITIES

The collection and use of data from various sources can provide important clues to law enforcement about an impeding major terror attack. Some of these data categories are listed below.

- **Motivational indicators:** religious fanaticism; supremacist ideology; pronounced antigovernment ideology; millenarian prophecy; paranoid conspiratorial worldview.

- **Organizational indicators:** attempts to selectively recruit members from the CBRNE-related sciences, such as veterinarians, doctors, physicists, chemists, biologists, and microbiologists, acquiring a high degree of technical and tactical sophistication; relatively long planning horizon associated with the use of a CBRNE weapon; highly disciplined, focused group which is "vertically" organized, highly integrated, and ideologically uniform, enabling its members to have the capacity to initiate and maintain in secrecy a large volume production line for CBRNE weapons.

- **Financial indicators:** large source of funding to procure materials and equipment, and provide training, facilities, and/or payment for personnel involved in the CBRNE project.

- **Logistical resources:** access to a relatively safe haven where development efforts can take place unhindered, or at least a location where these efforts can proceed with little chance of detection[22]; transnational network for acquiring raw materials and equipment; adequate capacity to transport the completed weapon to its destination.

- **Knowledge/skill indicators:** indigenous development of knowledge and skills within the group; necessary skills are obtained by utilizing experts from outside the group.

Early Warning Signs for B-Material Production

Special early warning signs have been identified for indicating the illicit production of B material by terrorists.

---

[22]It is possible to maintain a small-scale production of biological agents in an area as small as a sizable basement.

- **Production indicators:** purchase of biological cultures or toxins by a suspect group or individual; theft of seed stocks from hospital, university, or commercial laboratories; transfer of seed stocks from a nation-state-level biological weapons program; creation of a pathogen from genetic building blocks; purchase of typical equipment (fermenter, lyophilizer); information sought about obtaining samples of biological pathogens; requests made to civilian biological seed culture supply; evidence of castor bean plant(s), and/or crushed castor beans, or evidence of purchases in unusual quantities or evidence of cultivation; areas where agricultural or livestock disease outbreaks visited by individual suspect or group; shipments of supplies from a laboratory supply company(s) that included growth media; unexplained thefts of agricultural equipment (e.g. sprayers) or animals in the area; unusual interest noted in the acquisition of vaccines and medications; interest in specific locations of agricultural asset concentrations (maps, foreign requests for detailed data, foreign travel plans to rural areas or destinations of agricultural concentrations); noxious or unusual odors, not routinely associated with the area, around the suspect facility[23]; mechanical exhaust systems or air-filtration units, not consistent with routine building requirements, in the facility; unusually noisy fans audible outside a facility from oversized exhaust systems.

- **Indirect signs:** HAZMAT warning signs placed on tightly closed vials or containers; other HAZMAT signs on the material and equipment at the location; purchase of access to vaccines or antibiotics, such as streptomycin, tetracycline, penicillin, or ciprofloxacin; purchase or possession of unusual amounts of bleaching material, such as liquid or powder bleach, at any of the suspect facilities (such as apartment, house, garage); an improvised shower and eye bath located in an unusual area; textbooks, journals, and/or material discussing microbiology, biology, medicine, chemistry, explosives, poisons, and the like, including access to, or use of, dedicated online sources; display of unusual interest or surveillance of potential CBRNE targets, including obtaining maps, weather data, or other information covering specific civic areas, such as subways, airports, military and industrial complexes, or economic and agricultural areas.

- **Suspicious hardware:** laboratory glassware, hoses, or mortar and pestle in a nonlaboratory-type environment; agar plates, incubators, pressure cookers, centrifuges, fermenters, and/or mason jars; protective clothing, including surgical masks, gas masks, or self-contained breathing appa-

---

[23]The production of some biological agents may produce smells similar to a brewery or fermented grain.

ratus or respirators, rubber aprons, rubber gloves, and boots; fume hood, glove box, and/or HEPA filter; incinerator, incubator, large quantities of eggs, cell cultures, and/or small animals; lyophilizer and/or freeze dryer; small atomizers available, either stored empty or filled with biological agents.

- **Delivery system signature:** recruiting members or sympathizers who worked or received training in the field of aviation or agricultural sprayers; access to agricultural sprayer(s), commercially available for dissemination of material, including spray nozzles of varied sizes for aerosol dispersion; means to contaminate food or water supplies, or by spreading a contagious agent through physical contact; leaving unaccompanied packages in crowded or contained areas, such as in a train station, airport facility, subway station, supermarket, shopping mall, or department store; packages left near a ventilation system; pattern to the location and timing of "lost" packages.

Bioattack Indicators

Some indicators may point to an imminent or ongoing terror attack with B material, such as:

- **Behavioral indicators:** unscheduled spraying by aircraft, helicopters, or individuals, particularly at night; attempts to transport or abandon containers concealing chemicals; unusual or unscheduled window washing or power washing, especially when crowds are present; unscheduled presence of individuals in dust masks or other protective gear; tampering or unusual activity associated with water supplies, building ventilation systems, food supplies, or food distribution centers.

- **Epidemiological clues:** single case of disease caused by an uncommon agent (such as glanders, smallpox, viral hemorrhagic fever, inhalational or cutaneous anthrax) without adequate epidemiologic explanation; unusual, atypical, genetically engineered, or antiquated strain of an agent (or antibiotic-resistance pattern); higher morbidity and mortality in association with a common disease or syndrome, or failure of such patients to respond to usual therapy; unusual disease presentation (such as inhalation anthrax or pneumonic plague); disease with an unusual geographic or seasonal distribution (e.g. tularemia in a non-endemic area, influenza in the summer); stable endemic disease with an unexplained increase in incidence (like tularemia or plague); atypical disease transmission through aerosols, food, or water, in a mode suggesting deliberate sabotage (where there is no other possible physical explanation); several unusual or unexplained diseases coexisting in the same patient without any other explanation; illness that is unusual (or atypical) for a given population or age group (such as an outbreak of measles-like rash in adults); unusual pattern of death or illness among animals (which may be unexplained or attributed to an agent of bioterrorism) that precedes or accompanies illness or death in humans; unusual pattern of death or illness among humans (which may be unexplained or attributed to an agent of bioterrorism) that precedes or accompanies illness or death in animals; an unusually large number of ill persons who seek treatment at about the same time (point source with compressed epidemic curve); similar genetic type among agents isolated from temporally or spatially distinct sources; simultaneous clusters of similar illness in noncontiguous areas, domestic or foreign.

2.4. Legal Aspects of Emergency Situations[24]

2.4.1. THE IMPORTANCE OF LEGAL MEASURES DURING PRE- AND POST-ATTACK PERIODS

The idea that international diplomacy can control terrorism by itself is outdated in the new era of global terrorism. The threatening entities that societies face, and will face in the future, are not nation-states. They have their own resources from which to fund their actions and their own, increasingly virtual, constituents. Thus, in addition to diplomatic measures that were traditionally used against nation-state sponsors of terrorism, we must be able to employ a variety of social sanctions and various means to control terrorism. These should include legal measures, like law enforcement investigations, and judicial statutes against the forms of terrorism we know and will come to know.

One social structure that could be the most profoundly affected in the post-terrorism incident world is the legal system of the nation-state under attack. Fundamental changes, wrought by the act of terrorism, can occur, and occur quickly, in such social constructions as the law. For example, in the aftermath of significant warlike attacks, the USA has interned its own citizens and taken their property, changed its way of investigating criminal behavior, and otherwise invoked legal doctrines that seemingly undermine the fundamental character of the society.

Since the terror attacks in the USA on September 11, 2001, recognition of a potential concern over legal challenges—relative to emergency response to these types of events and the threat they represent to society— has

---

[24]Material regarding the US tort issues is based on a paper by Kristine Mullendore and James David Ballard delivered at the NATO STS-CNAD Workshop, Portugal, 2006.

emerged. In the aftermath of 9/11 first responder communities have seen an upsurge in the filing of lawsuits by plaintiffs seeking compensation for injuries suffered from both the attacks themselves and the public agency responses to them. These legal actions are based on liability theories deeply rooted in Anglo-American jurisprudence that allocate fault in an effort to determine liability, and then compensate for any damages suffered because of that fault in what could be described as a "corrective" justice approach.

It is axiomatic that any nation intent on preparing its modes of response in the fight against global terrorism needs to determine how its criminal justice system will be engaged in this fight, including determining whether or not law enforcement agencies will be employed to investigate, apprehend and detain, prosecute, and punish the terrorists involved in the execution and planning of these attacks (Kellman, 1999; Maogoto, 2005). Each nation should also anticipate that persons who are injured in terrorist attacks will seek to redress their harms in civil lawsuits, and that compensation will be sought for injuries measured in loss of life and property, and may even extend beyond those categories. The use of legal systems to resolve these civil law disputes, and to seek societal and individual justice by this process, carries a significant and troubling risk of disruption for societies that fail to plan for such contingencies.

This section examines the situations in which legal issues can and do impact public agency provision of emergency response services in the aftermath of an act of terrorism. This is done in an effort to determine how liability theories can impact fundamental professional service provision by emergency responders and security services in the aftermath of a terrorist event.

2.4.2. TERRORISM *VERSUS* CATASTROPHIC TERRORISM

There is no singular determinative definition of terrorism and perhaps few limits to the legal impact of current politics regarding the worldwide war on terrorism. In fact, the common theme present in discussions of terrorism is that it is something that is difficult to define, even though its acts are readily identifiable (Kingshott, 2003; Weinzierl, 2004). Despite the plethora of definitions about terrorism, there is also agreement among those who study terrorism that there are "shared characteristics" found in the varied forms of terrorist acts. Among these characteristics are:

- That, While the conduct involved will be dangerous to specific people and/or property, the fundamental goal behind these terrorist acts extends beyond whatever individual harms will result, and terrorism is used with the purpose of instilling fear in those who observe the act and its consequences.

- Instilling this fear in the general public is done with the purpose of causing some type of change in the behavior of the targeted individual, group, or nation.
- There is a strategic purpose in that a specific type of change is sought.
- The terrorist activity is organized and planned behavior.
- The terrorist actions break normal societal conventions of human behavior by targeting the innocent.

This list does not represent the sum total of factors relevant in determining whether an attack is defined as terrorism. The scale of the attack is also an important aspect in determining how society perceives and then responds to the threat or the actual event. For example, terrorism becomes *catastrophic terrorism* when the intended harms occur on a vastly larger scale that carries the potential of injuring thousands, if not tens of thousands, of victims (Kellman, 1999). Thus, it is not the terrorist's choice of technique, or method, which determines whether the event will be labeled catastrophic; terrorist actors could employ chemical, biological, radiological, or nuclear devices to achieve their desired level of social impact.

The intent that a terrorist attack be large-scale is manifested through the choice of weapon that is likely, under the specific circumstances, to result in a large number of victims, and can therefore be termed "weapons of mass victimization" (WMV) (Ballard and Mullendore, 2004). Still other commentators have noted that when the threat of terrorist attacks is compared with other current social problems facing nation-states (such as drug crimes or domestic violence), "terrorists have disproportionate powers over policy" (Bhoumik, 2005, p. 287). Thus, as these observers point out, the social impact of terrorism is larger than the immediate tragedy of loss of life and injury to the particular victims of the event.

Creating a social perception of a national security crisis is one of the strategic purposes in the choice of WMVs. They are chosen specifically to create social panic and social disruption in the normal functions of the targeted societies at the time of the attack, as well as during the aftermath of the mass destruction incident, when the targeted nations face challenges at many levels, including how their legal systems will respond. This challenge to nation-states' legal authority, or the underlying rule of law for a nation-state, can be as profound as the injuries and deaths that result from the actual attack. Policymakers should anticipate that WMVs will disrupt the customary functioning of the legal system, and should therefore recognize the need to engage in anticipatory planning as to how the nation's legal system should respond.

2.4.3. CRIMINAL JUSTICE RESPONSES

Threats to civil liberties that can occur during times of social upheaval have been specifically identified as being among the potential dangers that may damage the fabric of society. These "civil liberties" threats were part of the public discourse surrounding the Antiterrorism and Effective Death Penalty Act (AEDPA, 1996) that was enacted in the weeks following the Oklahoma City terrorism bombings in the USA (Ballard, 2005). The debate over AEDPA provisions resulted in the creation of a list of identified "terrorist organizations," criminal prohibitions against providing these organizations with any material support, and an expanded scope for deportation processes. These actions foreshadowed the sharp debate surrounding the post 9/11 passage of the USA PATRIOT Act of 2001. The passage of both these new laws created a high level of concern by observers as to how Congress should balance the legitimate national security concerns of the nation in contrast to protecting the civil rights of individuals in an age of global terrorism (Bhoumik, 2005; Guiora, 2005).

These challenges are especially difficult for democratic nations with strong social values supporting individual freedom, free speech, free association, and specific structural limits on the scope of governmental actions that may infringe on civil liberties. Some of these concerns include how to limit social responses to such tragedies, and what actions are available to combat terrorism when those responses interfere with rights of the criminally accused (Bhoumik, 2005). Thus, the very nature of democratic states contributes to their vulnerability to attack. These same democratic processes also make these nations attractive targets for political manipulation by placing social and political pressures on policymakers via the reactions of the threatened populace (Ballard, 2005). In this manner it can be argued that the threat of terrorism to the USA "comes not from the terrorism, but the response to terrorism" (Bhoumik, 2005, p. 309). The bottom line is that (a) democracy comes with a price in today's and tomorrow's world; and (b) democracy may be severely challenged as terrorists engage in deliberate attacks on the very foundations of that social order—the legal systems that are so justifiably venerated by such societies.

The USA PATRIOT Act's creation of "sneak and peek" warrants and secret inspections of library records can be seen as unfortunate, albeit necessary, intrusions into individual privacy. Citizens should embrace these civil liberty challenges to ensure the safety of a nation as it faces domestic and foreign threats of terrorist actions. On the other hand, these responses may inherently represent fatal changes to the character of the US society, destroying fundamental notions of civil liberties upon which it is was founded (Wattellier, 2004; Guiora, 2005).

The USA is not alone in facing these threats, though each nation-state has its unique concerns and must articulate these concerns within their democratic traditions. Israel, Russia, Spain, and India can all be placed on the list of nations who have experienced these types of attacks, and all of these nations have employed their functioning legal systems as an integral part of their national policy response to these attacks. (Guiora, 2005) These examples point to a maxim of modern social life: (a) there is every reason to believe that terrorist activities will be a continuing threat to the security of nation-states; and, by implication (b) nation-states need to specifically plan how to handle these challenges to their legal authority and processes (Bhoumik, 2005).

2.4.4. TORT LAW AS A SOCIETAL MECHANISM TO PROTECT THE INNOCENT INJURED

A fundamental tenet of US tort law is that where there is sufficient proof that someone has engaged in legally proscribed conduct, injured parties may have the courts order the wrongdoer to pay monetary damages as compensation for the injuries that resulted from that conduct. These legal proscriptions are based on the premise that society requires its members to exercise appropriate care and regard for other persons and their property interests. Those who fail to fulfill their socially imposed "duty of care" should then be financially accountable for the harms, rather than placing this burden on the innocent persons injured by their wrongful actions.

In those jurisdictions where these rules were based on British Common Law principles, there have been modifications caused by the evolving and developing jurisprudence, impacted by the USA's industrialized and capi-talistic economy (Berman and Reid, 1996). This is a "corrective justice" approach because of its requirement that the wrongdoer "correct" the harm incurred on the injured party by taking action "to restore the party to the pre-injury position" (Culhane, 2003, p. 1033).

The use of tort law to redress harms is embedded in US culture to such a degree that the USA is seen by the rest of the world as an overly litigious society; and, within the USA itself, this perception has resulted in many legislative initiatives to reform tort law at both the state and federal levels, which were referred to earlier (Mullenix, 2004). These reforms address a number of concerns by doing the following:

- Limiting the amount of recovery for noneconomic damages and elimi-nating punitive damages
- Reforming the determination of compensation where the collection of collateral source payments are involved

- Reforming laws governing the admission of expert testimony
- Modifying the role of the jury as fact finders in complex litigation
- Addressing the problem of frivolous lawsuits
- Limiting the extent of liability exposure
- Addressing the issues raised when litigants engage in forum shopping among possible legal environments present in a federalist country, and by the quality of judges that hear the cases (Mullinex, 2004).

However tort reform modifies the US legal system, though, it is important to note that these reforms do not undermine or alter the foundational legal principles. They leave tort law methodology intact as the appropriate model and process for recovery by allocating the burden of financial harm for the misconduct upon the responsible party.

Originally, the standard of care that was required of persons was that they act so as not to be either negligent or grossly negligent, or to knowingly or purposefully cause harm to others. Modern tort law developments in industrial society have created an additional area of liability with strict standards that impose liability on persons who produced goods to protect innocent users who have been harmed by the goods without regard to any fault in their choice of actions—engaging in the behavior is sufficient. Product liability law moved a considerable distance from traditional notions that financial responsibility should be taken away from the innocent injured and placed on a wrongdoer because of their "faulty" behavior. This shift away from "fault" created the context within which class action lawsuits and mass tort litigation became common litigation approaches for plaintiffs' lawyers, and these are the targets of much of the current objection to tort litigation that is being addressed by tort reform legislation.

One example of an alternative approach that was developed to address these concerns was when the American states started providing workers compensation programs as an alternative to tort lawsuits for employees injured in the workplace. This social policy protects innocent workers who have suffered injuries in the workplace because that environment is both controlled and created by the employer. These laws impose an affirmative obligation on employers to provide a safe work environment, or pay the costs if they do not; and as a part of a "corrective" justice approach to redressing harm, recognize the need to remove work-related actions from the tort law arena.

2.4.5. TORT LAWSUITS IN A POST 9/11 WORLD

Many tort lawsuits have been filed by the families of those killed or by persons injured by the 9/11 attacks on the World Trade Center and the Pentagon, by the flight that crashed in a field in Pennsylvania, by the first responders injured in the aftermath of these attacks or in the cleanup, and by persons present in the area who were exposed to the toxic environment in the aftermath (Hamblett, 2002; Ramirez, 2002; Mullenix and Stewart, 2002; Goldman, 2003; Neumeister, 2003; Mullenix, 2004). Among these lawsuits are ones brought against defendants who were also victimized in these attacks, and/or who were merely fulfilling a moral or legal obligation to try to assist those injured in these attacks. This latter category of civil actions revictimizes the "first responders," while also adding another layer of injury to society and to those harmed by these attacks. The net result is that these lawsuits, based on extraordinary circumstances, could undermine the fundamental theories of liability which are the foundation of civil law systems, thus helping to achieve the terrorist's goals. Thus, it is important for policymakers to address these concerns proactively and to mitigate potential harm prior to an attack, not *ex post facto* when the harms could be complicated and the debates could be harmful to the very foundations of democratic societies.

This effect of such legal challenges is not coincidental. It is foreseeable that fear is likely to destabilize a society under attack and—as a corollary intended effect—an even greater fear in the general public will be caused through catastrophic terror attacks (Bhoumik, 2005). The use of tort theories of recovery by innocent victims of these attacks to collect damages from other innocent victims, or from those who responded to the crises, undermines the US civil justice system, furthering the terrorists' ultimate goal of societal destabilization. US policymakers and lawmakers should fashion alternatives to the use of traditional "corrective" justice notions implicit in tort litigation by creating compensatory schemes similar to that created in the September 11 Victim Compensation Fund of 2001 (Fund), and codifying the conditions under which agencies, such as the FEMA, can address these injuries.

To continue the use of tort lawsuits as the appropriate method to redress the harms suffered by terrorist victims in a post 9/11 world fails to recognize the fundamental change in the nature of the risk of terrorist attacks faced by societal members. The corrective justice approach of tort law is not well suited to remedy the harms incurred in these situations, and its use for this purpose harms the US civil justice system, a system which is already under attack by tort reform measures motivated by intentions emerging from other unrelated aspects of social pressure.

Since the harms of a terrorist attack are readily anticipated, policy-makers and lawmakers should adopt a "distributive" justice system of no-fault recovery to provide compensation for those injured in catastrophic terror attacks (Culhane, 2003). Furthermore, such a no-fault system should also cover the first responders and bystanders injured by exposure to the toxic environments that these attacks can create. To date, these types of legal concerns have not been adequately addressed (Kellman, 1999; Steinhäusler and Edwards, 2005).

This area of policymaking and lawmaking is further complicated by the fact that "first responders" is not a monolithic category, but rather includes "fulltime employees, 'paid volunteers', organized volunteers, and convergent volunteers" (Steinhäusler and Edwards, 2005, p. 83). The duties and respon-sibilities of these first responders will vary widely depending upon their category and on where they are located on the hierarchical chain of command, as well as by the differences to be found within each state's laws. Yet, a distributive approach would benefit all such responders and allow for a more rational response to the social harm terrorists seek to impose on nation-states.

2.4.6. BREAKING THE MOLD

There are suits filed against al-Qaeda, as well as other individuals and governments that are believed to be responsible for planning and executing the attacks, which at first blush seem to be "corrective" in nature (Culhane, 2003). Here the plaintiffs are claiming that the governments named as defendants provided state support to the groups, and that establishes the jurisdictional basis for federal civil actions under the AEDPA (1996). Other cases base the liability of nations for the illegal actions of terrorists under customary international law, when the nation-state is the instigator of the illegal conduct and "is vicariously liable for every act of its own forces, of the members of its government, of private citizens and of aliens committed on its territory" if it neglects its duties (Maogoto, 2005, p. 264).

A nation's liability for the actions of individuals is complicated by the use of private contractors in the war on terrorism when these private contra-ctors engage in functions customarily the preserve of government officials. Behavior violating international standards also provides a "corrective" basis for recovery under the Alien Tort Claims Act and the Torture Victims Protection Act (Bina, 2005). However, other legal theories used to claim for recovery in the lawsuits filed after the terror attacks of 9/11 do not fit within this corrective justice model and, even where corrective justice theories would apply in redressing the injuries suffered by these victims, it is sugg-ested that the use of some type of victim fund that is funded and operated by the nation is a more appropriate social response where terrorism is involved.

Other lawsuits are based on customary legal theories of the duties of private individuals. These are applied to property owners, responders, and the airline industry, arguing that they should be held financially responsible for failing to prevent the attacks, failing to react appropriately in the aftermath of the attack, failing to take into account the toxicity of the environment created by the explosions, and/or to take necessary precautions for those who could foreseeably be exposed (Culhane, 2003). That these types of lawsuits would be filed as a result of the 9/11 attacks was so likely that Leo Boyle, president of the Association of Trial Lawyers of America, issued a statement shortly afterward, asking its members to enter into a moratorium against filing these suits for the first 6 months after the events occurred (Kay, 2001; Barkett, 2003; Mullenix, 2004).

2.4.7. DISTRIBUTIVE JUSTICE MODEL AND ITS LIMITATIONS

Given the current state of legal alternatives, and the messy ways they are being articulated in the courts, an alternative to tort litigation should be created under federal law for victims of WMV. Such an alternative system would start by implementing a distributive justice model, by which society makes the decision to provide a consistent, policy-based process that provides assistance to those harmed by terrorist attacks as well as other forms of mass victimizations such as natural disasters. The federal compensation fund, created after the 9/11 terrorist attacks within the Air Transportation and System Stabilization Act (ATSSA), could provide a model for this victim compensation "fund approach," because of both positive characteristics and the lessons it provides how not to construct such a Fund. There are many legitimate criticisms of this Fund, created within 3 days after the attacks, but it did include mechanisms to provide immediate assistance to the victims and families, while limiting other legal options (such as litigating claims in the civil justice system) if the victims accepted benefits under its terms (Mostaghel, 2001; Culhane, 2003). It is interesting to note that any possible culpability on behalf of the federal government was not considered within the terms of the Fund, and with sovereign immunity and other limitations against liability for discretionary acts of the federal and state government, there may be no need for such protections, although local governments would be at greater risk of liability. Hence, the proactive need for such liability protective mechanisms may well be employed to the advantage of the very foundational authorities for most first responder agencies, local governments.

The Fund also placed limits on the compensatory and punitive damages obtained by persons who elect not to make a claim under the Fund at an amount equal to the coverage provided by the insurance policies in effect at the time of the attack. One criticism of the Fund is that these limitations combine to effectively eliminate the possibility of tort recovery or compensation from the Fund for anyone suffering from latent injuries. To qualify for payment under the Fund, it is necessary to show that the claimant sought medical treatment within 72 hours of the attacks. When this temporal limitation is combined with the cap on recovery, this could mean that there may be no money available when the more latent injuries suffered by the first responders and local residents' exposure to toxic environments are disclosed (Culhane, 2003).

The Act that creates the Fund also provides immunity for persons who act to stop such terrorists in the execution of their plans, if the belief that a terrorist attack will, or was about to, occur is deemed reasonable, anticipating the desirability of social support for such actions with respect to any future terrorist activities involving airlines (Fund). This immunity seems to be aimed at allowing potential victims to react during the attacks, although it may also provide for more sinister applications when one considers how under duress of an imminent attack, law enforcement and other security forces may break the law to prevent an attack.

Another limitation on the Fund is that is does not cover victims who suffered only property injury (Culhane, 2003). This may well apply when the wider community suffers the financial burden of an attack versus only those who own businesses directly affected by the bombing or other terrorist tactic. A better response to catastrophic events would both recognize the risk of the future use of WMV, and anticipate that there will be innocent victims who may, without such a fund, sue other innocent victims to receive the financial support they need to compensate for such losses. This better response would be to have the legislation in place before the event occurs and have it address such contingencies.

The financial effect for victims and first responders immediately after a mass casualty incident can be devastating. Traditional accounting and actuarial projections have difficulties even calculating the probabilities and risks of mass casualties from terrorist attacks, because they lack the necessary factual information to make these calculations and risk assessments—which are termed low-probability/high-consequence events. Insurance policies are a widely used method to distribute risks of harm throughout a social group. These policies effectively dilute the impact of the injury on the individual. The inability to determine risk makes insuring against these types of attacks with traditional insurance policies a questionable business practice (Manns, 2003). The inability to purchase insurance further increases the financial risk for those who are sued. These types of concerns, and the interests of landowners and developers implicated by this situation, had enough political clout to result in Congressional enactment of the Terrorism Risk Insurance Act of 2002. The concerns of individual victims of WMV with personal injuries are no less important.

2.4.8. NOT UNUSUAL POLICY

Congress has previously used a fund approach as a national response mechanism to address tort claims. The compensation plans used in the Dalkon Shield and Agent Orange cases demonstrate the feasibility of that type of managed administrative response as a replacement for mass tort litigation. In those instances the plans addressed liability issues for injuries occurring and/or discovered both before and after the plans were created.

Other policy-based response mechanisms previously used include the Price Anderson Act's provisions for handling tort liability that should arise from potential nuclear accidents, the National Childhood Vaccine Injury Act of 1986, and the Superfund that was created for environmental site cleanup (Floering, 2002).

These Acts are not without their own issues, problems, and critics. For example, Berkovitz (1989) notes that the Price Anderson Act has been criticized since it was not originally intended as a model compensation system; nor was it fully functional after amendment. When amended in 1988, the Price Anderson Act revised upwards the amounts of compensation, thus providing evidence that such periodic adjustments are necessary for compensation programs to stay current. Lastly, the original Price Anderson Act offered protection to victims, but equally as important, and subject to some criticisms, were the protections afforded to those who could be found liable for serious accidents involving nuclear and radioactive materials.

2.4.9. FUND REPLACEMENT OF TORT ACTIONS FOR VICTIMS OF WMVS

It is not possible to anticipate every type of tort claim that could arise from a terrorist attack. Furthermore, litigation may still occur under tort theories, even with a preemptive no-fault Fund mechanism, since these policies should not be intended to completely replace the use of courts. Therefore, some of these tort-based claims can and should be anticipated, and appropriate alternatives developed. The potential claims based on violation of real property rights should be included in the Fund's scope, as well as any claims brought by convergent volunteers, and those that would otherwise be workers' compensation cases, as well as those that could be brought by innocent bystanders, such as those injured by the toxic environments created either by the attack or the response to the attack (Reynolds, 1996).

That civil actions are rarely resolved by trials, as the vast majority of them are settled out of court, also argues for the use of a no-fault Fund to resolve these disputes. Plaintiff's lawyers should not be managing the distribution of resources to victims, where there are models for using administrators with no financial stake in the resolution of the claims, possibly similar to federal special masters, who would be more appropriate choices to obtain distributive justice (Ramirez, 2002; Mullenix, 2004).

A visionary approach is needed here. The world post 9/11 is not the same place it was on September 10, 2001, and US lawmakers not only should recognize that it is a changed world, but also recognize that there are changed governmental responsibilities in the administration of civil justice. Societies using tort theories of recovery operating under the rule of law cannot expect this model to effect an equitable distribution of the resources that are needed to redress injuries that are inflicted by persons who reject the notions that underlie the rule of law. Where terrorist's actions are involved—especially those involving catastrophic terrorism—resolution of the claims of innocent victims should not be left to traditional civil law "corrective" approach.

2.4.10. POLICY CORRECTION NEEDED

The September 11 attacks were not the first terrorist incidents signaling the need for lawmakers to develop a systematic and equitable approach to redress the injuries of victims of terrorist actions. During the 1980s financial support was provided to the Iranian hostage victims after their release under the Hostage Relief Act of 1980 (Mostaghel, 2001). At that time, prior to the Congressional enactment of the Relief Act, the President's Commission on Hostage Compensation recommended against the use of a tort law approach to compensate these victims, and suggested that damages, such as those provided to prisoners of war, was the appropriate response. The Victims of Terrorism Compensation Act, part of the Omnibus Diplomatic Security and Antiterrorism Act of 1986, was created as a legislative recognition that persons who engage in the work of the nation should, as well as their families, be compensated for injury and death (Mostaghel, 2004). Legislation was passed that provided compensation in 1990 for the victims of Pan Am Flight 103 over Lockerbie, Scotland; in 1996 for the victims of the Oklahoma City bombing, and in 2001 for the September 11 attack (Mostaghel, 2004; Wieentge, 2005). The USA, as a nation, has started to use mechanisms that approximate those suggested herein, but a more formalized means for terrorist consequence compensation should be put into place.

The pattern is evident—the need for this type of financial support is a recurring one. When responsive and reactive action is taken to a particular incident, the remedy has a high potential to result in disparate treatment of victims, and a failure to address the unique needs of the first responder communities. Legislation is, by its nature, meant to be general in application and prospective in coverage, while tort cases are designed to resolve previously existing disputes between private individuals and wrongdoers. Using particularized reactive legislation to redress the injuries of victims of terrorism on a case-by-case basis will necessarily result in inequities for the victims, and the use of tort law will inevitably result in the revictimization of those who are on the front line acting to protect all of society.

It is evident that it is no longer September 10, 2001, and the legal mechanisms of that world need to be evaluated and altered to reflect that change. Congress should develop a catastrophic terrorism victim fund that will rectify the potential for these inequities and provide necessary support to those on the first line of response. The serious questions of justice and fairness created by previous *ad hoc*, after the fact, policy responses should be corrected. After noting the inappropriateness of using tort law in these cases and detailing his criticisms of the payments provided under the Fund created for the 9/11 victims under either a corrective or distributive justice approach, Culhane (2003) states it well when he concludes: "The better approach would be to view injury and death from terrorism as social risks, offering basic compensation to those who need it, while undertaking on-going and parallel efforts to address extreme distributional inequities" (p. 1107). It is a social mandate that a nation protects the first responder community from liability claims and potential financial harm for their service to society.

2.4.11. LEGAL ISSUES FOR EMERGENCY MANAGEMENT PLANNING

Alerting and Warning for Terrorism[25]

Public information campaigns are an important tool to reduce the impact of a terror attack on members of the public, by alerting them to increased threats in advance of an attack. Such a campaign should address the timing of possible attacks, the definition of the target audience, as well as the specifics of appropriate responses for citizens at home, at school, and at the

---

[25]For a discussion of alerting and warning mechanisms and systems, see Chapter 3.

workplace.[26] However, it should be noted that such an information campaign may have a legalistic demand. In efforts to be proactive and limit legal liability, governments should consider how such campaigns inoculate them from potential risks of legal actions when victims claim their appointed political authorities failed to protect them from terrorists.

Technical alarm systems range from traditional technologies (sirens, police car public address systems) to new technologies (self-registration for notification, police-based crime alert e-mail, cell phone, BlackBerry, self-turn-on radio and television). The balance between protecting the public interest and limiting civil liberties is an issue confronting public agencies in the new war on terrorism. In part, this dilemma rests on legalistic issues. For example, if government agencies fail to warn the public of an impending or potential attack, they may be held responsible for that failure to inform. The forum wherein the information is presented may likewise be problematic given sensitivities to those with disabilities and/or limited means to access the various technologies that could be employed in the notification process.

The government has the responsibility in many cases to keep their citizens informed in the event of an accident, tragedy, disaster, and/or attack. This is an issue that is both pragmatic and mandated. Pragmatically the governmental entity should seek to normalize social interactions as quickly as possible and to reestablish authority, control, and some sense of normalcy. The mandate is that if they do not do this "normalization," the actual authority given the political entity may be undermined and could even lead to a change in power structures.

Evacuation and Shelter in Place

The orders for evacuation and/or to shelter in place during a significant terrorist attack come from the legal authorities in place at the time of the incident. For example, many forms of emergency declarations exist in various jurisdictions, and these would be invoked in the event of a large-scale incident. The National Response Plan and the Stafford Act, in the USA, define which levels of government have which emergency powers, while the US Constitution acknowledges that all disasters are local.[27] It is

---

[26]For examples of such campaign materials, see "American Red Cross Homeland Security Advisory System Recommendations for Individuals, Families, Neighborhoods, Schools and Businesses" at http://www.redcross.org/services/disaster/beprepared/hsas.html.

[27]Article IV, Section 4, guarantees the states a republican form of government and protects them against invasions, and "on application of the legislature or the executive ... against domestic violence." Amendments 9 and 10 reserve all powers not assigned to the federal government to the people and the states. The Constitution is silent on disaster

the responsibility of the local government to declare a local emergency, and of the governor to declare a state of emergency, and to request a presidential disaster declaration. Evacuation orders in the USA must come from local or state authorities.[28]

A critical issue is how to accomplish post-event evacuation notifications, since this involves technical topics, social systems, as well as national and international media. The legal authorities for such notifications should be in place prior to the event, and prescriptive measures detailed in those regulations, statutes, and laws. Periodic review of these legal authorities, necessitated by changing technology, should likewise be mandated in the legal authorities.

Cross-Border Interoperability Across Systems

Preexisting cooperation agreements should be negotiated between response partners if they wish to avoid logistical, operational, and/or legalistic challenges in the aftermath of an event. Logistically, these mutual aid agreements should predetermine, as much as possible, who will run the emergency response, what the roles are for those who come to aid in the response, and so on. In some cases exemptions to existing laws and regulations may be necessary in the event of a catastrophic event. For example, if a country has specific regulations on medical practice, the use of doctors from other countries during a crisis may require an exemption to the medical licensing process.

Likewise, border security and custom regulations may need to formalize exemptions for certain response technologies and/or supplies that teams responding to an event may wish to bring into the affected country. Weapons are currently the area of greatest concern, as many nations have prohibitions against the importation of firearms. Treaties for mutual defense against war attacks have empowered responding military units to bring their full complement of equipment with them, without reference to local firearms regulations. Such protections do not extend to law enforcement mutual aid following a disaster. The event of 9/11 demonstrated the many types of law enforcement-related assistance that may be needed after a catastrophic disaster. Law enforcement officials would be reluctant to enter a disaster zone without their side arms and other weapons, since part of their role could be crowd control, evidence protection, scene security, and first responder force protection against secondary attacks by the terrorists.

---

response, but clear on preventing federal intervention without the request of the executive or legislative branches of state government.

[28]Note the discussion of Hurricane Katrina in Appendix 5.

Response

Search, rescue, and recovery operations after a catastrophic incident of any kind, natural disaster or human initiated event, can induce legalistic questions and dilemmas. Under the circumstances, without normalized judicial processes, and perhaps in the absence of local community-based police services, emergency responders must be empowered to make quick and liability-free decisions regarding fundamental questions such as entry to structures without a warrant or authority to do so, access to records and information without legal restrictions or those set by precedent, and any number of other legalistic issues that may arise under such circumstances. Pre-identification of legal issues and development of internationally agreed-upon response protections would hasten the response and prevent post-event legal arguments.

Corpse Identification

In the aftermath of a terrorist attack the resulting corpses may be evidence, may contain significant clues to the attacks and its perpetrators, and must be treated by public authorities as a significant piece of evidence for investigators. This complicates legal issues associated with identification, since the corpse may need to be kept for forensic specialists to examine prior to being sent for burial, cremation, or similar services. The collection of DNA from the same bodies may likewise offer legal challenges, if the religious preferences of the victim and/or his/her family are considered.

In the aftermath of a terrorism incident two important things transpire in addition to assisting the victims and helping the community recover: (a) investigating the crime that the terrorism act represents; (b) some form of judicial process wherein the evidence that is collected at the crime scene will be presented. In some cases these two acts contradict each other or may be at cross-purposes in the field. Investigators may be trying to prevent the next attack, and evidence preservation may not be their first priority at that time. Those who need to protect the chain of evidence may, by necessity, impede the investigation of the crime. The critical item to remember is that a terrorism scene is a crime scene and that the victims, building, airplane, vessel, or whatever the location and outcome of the attack are evidence. Sensitizing everyone involved to this reality will assist in the legal processes that follow, be they investigative or judicial. This will be particularly critical if cross-border issues in evidence collection and case preparation need to be resolved.

Crucial issues are due process and equal protection standards, which are far from uniform across NATO. Historic documents based in British Common Law, Napoleonic Code, Roman Law, and many other evolutions of jurisprudence govern each nation during domestic tranquility. A cross-border terrorism response plan must include a consensus on the definition of due process for anyone suspected of being involved with the terror attack, on the concept of chain of evidence, and on the protection due to the person apprehended. Consensus must be reached in advance on definition of torture, on acceptable investigative methods, and on criminal procedures that are acceptable to the NATO alliance partners.

An examination of the management of the apprehensions of the Madrid bombers and the London bombers shows some of the differences in police procedures caused by basic jurisprudence in each state. The cross-border terrorism response plan might empower Interpol as the lead law enforcement agency, and rely on the standards currently in use in drug interdiction and other international criminal cases. Whatever standard is agreed to will transgress the sovereignty of the nation-state in which the terrorist disaster occurs. Decisions on how and when that sovereignty is infringed must be made in advance of any event, when calm discussions and weighing of various claims can be undertaken in a collaborative environment.

3. PROTECTION AGAINST TERRORISM THROUGH EMERGENCY MANAGEMENT STRUCTURES

The liberal democracies of the world are fighting a war with terrorists who are using fourth-generation tactics.[1] Collective action across borders against terrorists may be the most effective weapon available to the nations striving to prevent terrorist acts on their soil. The response from the nations has to be effective, yet meet the standards of the international conventions and treaties. As described in Chapter 2, law enforcement entities are organized to use their skills to track, counter, arrest, and convict terrorists. Because the goal of law enforcement is prevention of such acts, their investigations and preventive measures may run counter to traditional notions of civil rights in an open society. Terrorism poses a challenge to protect the residents' physical safety and their civil rights at the same time.

Emergency management structures in each nation, and within the international community, are tasked with protecting the potential victims of terror, and providing immediate life-saving response to an attack. Population protection against terrorism has a great deal in common with population protection against any disaster. An explosives attack against a population has similar consequences, whether it was intentional or accidental, whether it was caused by international terrorists or a disgruntled worker. A chemical attack and a chemical accident require similar responses for population protection, as do radiological events. Biological events will surely be treated as naturally occurring until the epidemiology and forensics define an intentional source. Only nuclear events will clearly be intentionally caused.

One of the challenges for emergency management is to organize its resources and activities to achieve the most effective response to terrorism. This sector's activities include all four phases of emergency management: mitigation, planning/preparedness, response and recovery.[2] Previous works in this series have covered the mitigation and preparedness aspects of the international response to the terrorist threat (Steinhäusler and Edwards, 2005). The foci of this review of emergency management structures are the requirements for protecting the population when an attack is imminent, and responding to the scene of a terror event.

---

[1]See Appendix 2 for a complete description of the Generation Warfare theory.
[2]The "four phases of emergency management were first defined by the National Governor's Association in 1979. See National Governor's Association, 1979, *Emergency Preparedness Project Final Report*. Washington, DC: US Government Printing Office.

F. L. Edwards and F. Steinhäusler (eds.), NATO and Terrorism, 49–82.
© 2007 *Springer*.

3.1. Structures for Managing Alerting and Warning for Terrorism

One of the most effective ways to protect the civilian population of a society against terrorism is to have an effective system of alerting and warning. While educational information needs to be repeated frequently (Lopes, 2002) the warning message must be issued in a timely fashion, but not so often that the population develops "inattentional blindness" and ceases listening to the protective messages from the government (Vedantam, 2006). In addition, the technology used to deliver the warning must be effective and appropriate to the community's lifestyle and needs.

3.1.1. PUBLIC ALERTING AND WARNING

The target audience for the message must also be carefully considered in designing the content (Lopes, 2002). A warning aimed at the entire society for a generalized threat would be issued in different modes than a warning to a neighborhood for a specific threat against a critical infrastructure. While the electronic media may be appropriate for generalized informational announcements, targeted delivery with more immediacy is needed for neighborhood scale disaster threat notification. Door-to-door delivery of fliers, telephone notification systems, and public safety vehicle loudspeaker systems will have more effect in a neighborhood.

A crucial difference between the generalized warnings and the neighborhood-specific warnings is that the content can be customized for the best reception by the neighborhood. Materials can be delivered in appropriate languages other than English. They can rely on the use of culturally competent delivery methods that respect cultural norms regarding age, gender, and trust. Rocky Lopes points out that people do not want to believe that they are at risk, so the more specific the warning is, the more likely it is that the warning message will overcome the denial generally experienced by individuals (Lopes, 2002).

Information management occurs in several steps. First the information must be coordinated with the law enforcement community to ensure that the information is accurate, timely, and does not endanger an ongoing investigation. Second, the content has to be developed to ensure that it is understood by its target audience, and motivates them to act (Lopes, 2002). The challenge of public information about terrorism is to find the golden mean between the need to repeat educational information multiple times for it to be received and assimilated by the hearer, and the auditory and mental fatigue that arises from hearing the same message or seeing the same poster every day.

An additional complication is the difference between natural disaster warnings and terrorism warnings: the causation factor. Amy Ding points out that warnings of natural disaster have no impact on the outcome of the event. A flood will occur wherever it occurs based on topography and meteorological conditions, regardless of whether people are told about it in advance or not. Conversely, the warning of a terrorist attack may cause the attackers to change their plans, switch locations or delay the attack (Ding, 2006). Therefore, while the public assumes that natural hazard predictions are based on science, and have a measurable probability of occurring, their experience with the terrorism warnings is that they are not reliable. It is also hard for the public to understand that while terrorists threaten the nations every day, the work of the international and national law enforcement agencies results in interdictions or diversions in many cases, so that threats are neutralized. To protect their sources the agencies often do not announce their successes, reinforcing the "cry wolf"[3] attitude of the public. As with all crime, it is hard to measure the effect on the public of the crime that was not committed.

Public information content has two elements. The first is the education of the community so they know what to do when they receive the message to take action against an imminent terrorist attack. The second is the warning of immediate threatening activity. For example, during World War II people received public information about preparing shelters in their homes, stocked with food and water, which would protect them against potential bombings. When the enemy planes were in the air, a siren system was used to alert the public to act on their pre-event education and go to the shelters that they had already created.

Similarly, effective public alerting and warning against terrorism requires that the public is first educated about steps to take to enhance their chances of surviving a terrorist attack. The Department of Homeland Security, the American Red Cross and numerous other public and private agencies have prepared publications outlining the threat, the steps to take to prepare for an attack, and the steps to take during an attack.[4] Materials define how to prepare at home, at school and at work for various types of terrorist attacks.

---

[3]Aesop's Fables, "The Boy Who Cried Wolf". This is an ancient tale of a shepherd boy who cried "the wolf is chasing the sheep" because he was bored. The villagers tired of responding to his false alarms, and when the day finally came that the wolf did come, they did not believe his cry for help, and the wolf triumphed. Many members of the public think that the government is falsely "crying wolf" in a desire to divert attention from something else.

[4]For example, The American Red Cross, 2006, *Talking About Disaster: Guide for Standard Messages*. Washington, DC: http://www.redcross.org/disaster/disasterguide/standardmsg.html; and Homeownership Alliance and Department of Homeland Security, 2005, Emergency Preparedness Guide: Protecting Your Home and Family. National Association of Realtors. http://www.realtor.org/GAPublic.nsf/pages/emergencypreparednessguide?OpenDocument.

Specific actions include the development of "safe rooms" within the home where chemicals can be excluded, creation of family plans for reunification during a disaster, and stockpiling of essential supplies for use during and after a terrorist attack.

The US Department of Homeland Security developed a "color code" system for use in informing the public about the level of threat from terrorist that the nation was experiencing. This action was based on Homeland Security Presidential Directive-3 that required creation of a "Homeland Security Advisory System to provide a comprehensive and effective means to disseminate information regarding the risk of terrorist acts to Federal, State, and local authorities and to the American people." The most dangerous period is red, then orange, yellow, blue, with green being "low risk of terrorist attacks" (Bush, 2002).

However, since the color code's inception no period has been less than yellow, which is "Elevated: Significant Risk of Terrorist Attacks" (Bush, 2002). The result is that yellow has become the "new normal" for people. In a study of families living in Chicago, Amy Ding modeled their behavior and discovered that frequently issued yellow alerts resulted in decreasing response by the families, until they no longer responded to the announcement of threat levels at all (Ding, 2006). Therefore, public officials must measure their use of threat warnings. Paul Light, a professor at New York University who studies government reform, notes that, "A warning is only effective if people know the action to take—and the federal government offers sound advice. ... The federal color-coded terrorist alert system, meanwhile, has had a numbing effect because it's not prescriptive" (Simon, 2004).

3.1.2. PUBLIC WARNING TECHNICAL SYSTEMS

Public warning systems have been in place since the days of the town crier and the night watchman. Verbal messages and alarm bells have long been a fixture of urban life. In the twentieth century the advent of aerial bombing led to the development of audible siren systems that were radio controlled by a central authority to warn the public of impending attack. The warning had to be loud enough to be effective, and delivered enough in advance to allow people to go to prepared shelters. In the 1980s, with the end of the Cold War, the US government stopped funding the national siren system, and most were dismantled because the cost of maintenance was too much for local governments. For example, in the 1990s Washington, DC dismantled its 100-sirens system, leaving just a few at George Washington University and Bolling Air Force Base (Associated Press, 2004).

Terrorism has provided a new impetus for public warning. The sirens filled a need for notifying people who were outdoors and away from radios and televisions. San Francisco, California, has installed new sirens in its Union Square area and throughout the community that are capable of tone alerts and voice message broadcasts (72hours.org). They plan to use the sirens for large population outdoor events, such as Halloween in the Castro District, but their real focus is the ability to warn people outdoors of impending disasters.

In 1963 the Emergency Broadcast System (EBS) was developed in the USA to allow messages from the President about imminent attack to preempt any radio or television programming (Federation of American Scientists, 2007). There was a key radio station in each jurisdiction that received the Presidential interruption on a special system. They then used a special audible code to indicate that this was an emergency warning message, to draw attention of the listener to the importance of the upcoming message. In the 1970s when the FEMA adopted an "all hazards approach" to emergency management the EBS began to carry weather messages and other government warnings. The system never delivered the nuclear threat message for which it was designed, but it did deliver over 20,000 messages between 1976 and 1996 about civil emergencies and warnings of severe weather hazards (Federation of American Scientists, 2007).

In 1997 the EBS was changed to the Emergency Alert System. The technology changed to digital, and all FCC licensees became partners, including cable systems with 10,000 subscribers or more. The technology is the same as that used by the National Weather Service for its weather radios, so information can be more readily shared. Advances in technology continue to make this system more functional as an all hazards warning system. Also, specially equipped consumer products, such as televisions, radios, pagers, and other devices, can decode EAS messages. The consumer can program these products to "turn themselves on" for the messages they want to receive (Federal Communications Commission).

National Weather Radio has long been a staple of tornado and hurricane-prone communities. The radios are silent until there is a weather-related disaster warning for the area in which they are located. Using Specific Alert Message Encoding (SAME) the radio issues an audible alarm followed by a message from the National Weather Service defining the problem and the probable area of effect (NOAA, 2005). These radios are inexpensive, less than $50 per unit in 2007 (NOAA, 2007), and silent until a local disaster warning is issued. However, the Partnership for Public Warning estimates that there are no more than 15% of the homes in America that have weather radios (Associated Press, 2004).

A final method long used for public warning is the public address system on a police car or fire engine. The loudspeaker system allows the driver to communicate a message to a community as he drives along. San José, California, routinely uses this method for flood warnings in low-lying neighborhoods. The technology is used every day for other public safety purposes, so it is always in good working order, and the drivers are familiar with its operation. Further, the message can be customized for each area, and a bilingual driver can be assigned to neighborhoods that need explanations in a language other than English. It is easy for people to walk up to the vehicle and ask the driver for more details. The drawback is that the information is heard only for a block or so at a time. Also, people inside homes may not be able to hear the message. Although drivers use the vehicle's emergency siren to draw attention to their presence before beginning the announcement, people living in areas with frequent emergency vehicles visits may not be adequately "alerted" by yet another siren.

So-called "Reverse 9-1-1" systems have been proposed as another warning method (Rein, 2006). These systems deliver a recorded voice message to phones in a given area based on a computer program. While these may work adequately in small areas or small communities, the system is limited to several hundred calls per hour. In a city with hundreds of thousands of households, the system would be too slow to reach all of the at risk households. The length of the message further slows down the delivery, so generally messages are limited to a few sentences, usually directin people to their radios for more details. If the person who answers the call does not speak English, is hearing impaired, or has disabilities that prevent them from acting on their own, the one-way message may not be useful or actionable for them.

Since the end of the Cold War there has been a boom in technology for various types of communications. The ubiquitous presence of computers, cell phones, pagers and BlackBerrys has encouraged public safety agencies to try various methods for public warning that rely on these new consumer-owned technologies. For example, the San José Police Department offered a crime alert e-mail system to community members who signed up on their website. Participation in the program was voluntary, and the consumer was responsible to maintain the contact when his technologies or addresses change. Washington, DC, also had a community-drive alert system. Only 18,104 people had signed up for the disaster alerts by January 2006, 18 months after it was introduced (Rein, 2006). The Washington system also provided for notification through any combination of phones, pagers, computers, and other systems that the consumer registered.

High-tech methods may not appeal to all community members. Older people and poor people are less likely to own communications technology beyond a home phone. The very young and the very old are unlikely to be well served by technology based communications systems. People for whom the dominant national language is their second language will also struggle with verbal warnings. Reynold Hoover of FEMA has noted that the challenge is to begin with public education to make the warnings actionable. "Part of our challenge is to educate a whole new generation of Americans about what to do if it were an actual warning" (Associated Press, 2004).

The Partnership for Public Warning has expressed concerns about the slow pace at which the federal government is addressing the problems of public warning. They have suggested the development of radios and televisions that could turn themselves on to broadcast warnings, better integration of warnings at the federal level, and development of a national warning day when local governments would educate their residents about how to respond to the threats to their communities (Associated Press, 2004).

Before its dissolution in 2005 the Partnership for Public Warning created "The Common Alerting Protocol: An Open Standard for Interoperability in All-Hazard Warning," (Partnership for Public Warning). The goal of the Common Alerting Protocol was to create a method for warning dissemination that could be used on multiple technologies to integrate the warning activity. While this is a step toward better integrating the delivery of the message, there are many challenges facing the emergency management community as it responds to warning the community effectively about terrorism and other disasters.

3.1.3. POST-EVENT MESSAGES

Post-event messages may be delivered by technology or by social systems. The technologies are the same as those used for verbal warning messages. Electronic media will be the primary source of information for most community members. In addition to describing the event, and thereby providing situational awareness for the community, the electronic media in a community will try to interview key community leaders, such as the mayor, police chief or fire chief. Internet-based technologies, cellular phone-based technologies and text messages delivered through phones or pagers may be used by local government agencies to provide residents with information on shelter locations, boil water orders, immunization availability, and other essential information.

Social systems may also be used to deliver post event messages to community members. These include neighborhood-based telephone trees, affinity groups like the Parent Teacher Association or a church group, and social structures like businesses. The information that they relay could include information on schools that are open and closed, business operations, and the location of FEMA Disaster Application Centers and Small Business Administration Disaster Centers in communities.

Often a post-disaster community is without power, making some of the technologies fail to function. Even battery operated cellular phones and pagers will eventually have to be recharged or die. In such situations runners, handwritten notices and posters, and community meetings become essential ways to communicate among the residents. Emergency plans should including poster paper, marking pens, and wipe-off boards to use when "low tech" solutions provide the best community information exchange.

Post-event messages often are based on assumptions of community preparedness. In the USA the federal government has adopted the Community Emergency Response Team (CERT) model of neighborhood organization (Department of Homeland Security). The goal is to train neighborhood members how to care for themselves and their neighbors following a disaster, regardless of its cause. Training includes home and personal preparedness for the risks of that community, fire suppression and household HAZMAT management, Simple Triage and Rapid Treatment (START) approach to medical care, and light search and rescue. Government leaders assume that community members, using these skills, will enhance the capabilities of the first responders, and help their less injured neighbors. CERT recognizes that professional first responders will be committed to managing HAZMAT releases, collapsed buildings, and conflagrations resulting from the attack. Their focus will be on the site of an attack and the hundreds of patients at that location. The question is whether the impacted community will have enough trained CERT members to fulfill the community needs.

Israel also has training for community members in post-disaster self-sufficiency. During the first Gulf War the government provided families with gas masks and information on creating safe room within their homes, in case of chemical attack. Because so much of Israel is within range of Scud missiles that could be carrying chemical warheads, new homes in key areas of the country must be built with "safe rooms" designed to Israeli Defense Force standards.[5] In addition, most Israeli citizens over the age of 18 years have had military training, so they know basic first aid and understand the mechanisms of injury from weapons of mass destruction.

In nations with similar training programs there are assumptions that the public education messages have been understood and remembered. These assumptions may be incorrect. In nations that did not perceive a high terrorist threat, public alerting and warning of a terrorist event may have to include a significant educational element that is delivered "just in time," to garner the

---

[5]For more information, see the Israeli section of Chapter 4.

needed citizen response. Websites, Internet messages and the electronic media may become the best methods for providing essential information to the residents.

3.1.4. ROLE OF GOVERNMENT VERSUS RESPONSIBILITY OF INDIVIDUAL

At what point does the individual become responsible for his own safety? Educational material is readily available at a variety of websites and from all levels of government. Local governments offer training classes and business preparedness conferences. Schools, hospitals and critical infrastructure industries have special regulations and training for responding to disasters including terrorism. The constitutions of most democracies limit the level of intrusion by the government into a person's life. Government can suggest and advise, but except for a few critical resources, the owner or the resident is responsible for himself and his own decisions. As we saw in Chapter 2, these constitutional and legal protections can result in a community that is not self-reliant, and expect government care after the disaster.

Hurricane Katrina in New Orleans, Louisiana, is often used as the paradigm for government's failure to respond adequately to the post-disaster needs of the community. Yet most of those needs were created because people made the independent judgment to ignore the evacuation order and stay in their homes or in their neighborhoods regardless of the impending storm. They had full knowledge that the storm was coming, yet they chose to ignore the evacuation order and stay in the city. The result was that not only were thousands of lives lost, but hundreds of first responders had to risk their lives wading through contaminated flood water to do a door-to-door search for stranded residents. Chapter 4 contains detailed information about that search and rescue effort. Appendix 6 contains a critique of the response.

3.2. Emergency Management Structures for Managing Response

Protection of the safety of residents is one of the primary motivations for the creation of nation-states. Most national constitutions place the requirement for maintenance of public safety and the prevention of invasion at the top of their organic documents. For example, the Constitution of the USA begins, "We the people of the United States, in order to form a more perfect union, establish justice, insure domestic tranquility, provide for the common defense, promote the general welfare, and secure the blessings of liberty to ourselves and our posterity, do ordain and establish this Constitution for the United States of America."

3.2.1. LOCAL, STATE AND FEDERAL RESPONSE STRUCTURES

Modern states have professional law enforcement and fire officials to oversee the safety of communities. They are joined by professional emergency
managers in planning for the maintenance of public safety. As was discussed
in Chapter 2, the risks to communities have been evaluated, and appropriate
levels of training and equipment are planned for, and provided as far as the
budget permits. In some nations the fire service is an all-volunteer force, as
in Austria, and volunteer departments predominate in some states of the
USA. In some nations the police force may be a national organization.

Regardless of the organizing mechanism, public safety is one important
focus of government at the local, regional/state and federal/national levels
for all of the members of NATO. In the past there had been a clear distinction between domestic disturbances, which are handled by the "first responder" services of fire and law enforcement; and the attacks by other state
entities, which are handled by the military services. Terrorism has introduced
a hybrid attack that has domestic impacts but may have a foreign origin.

In 1996 the USA because preparing its first responders for weapons
if mass destruction attacks. The Nunn-Lugar-Domenici legislation created
the Metropolitan Medical Strike Team (MMST) system to train and equip
first responders for terrorist attacks with weapons of mass destruction
(WMD): nuclear, biological, and chemical (NBC) weapons (Winslow, 2001).
The programs began with the 25 largest cities, along with Honolulu and
Anchorage because of their isolation and perceived risk. By 1999 the
program had grown to the 120 largest cities and two regions: New England
and the Texas Border. The goal was to ensure that first responders in a city
under attack could protect themselves and save the lives of the victims. The
bombing of the Oklahoma City Murrah Building in 1995 had demonstrated
that a city's local first responders were the only professionals close enough
to the victims to save their lives. For example, no living victims were rescued
from the Murrah Building after the first 12 hours. All live saves were accomplished by the Oklahoma City Fire Department and their immediate mutual
aid partners (The City of Oklahoma City, 1996).

The MMST program provided for six federal partner agencies to assist
the cities in preparing for possible terrorist attacks. The Department of Health
and Human Services provided a project officer to guide the planning effort.
The Department of Defense provided training in WMD response for fire,
law, HAZMAT and emergency medical services personnel. The FBI provided an adviser to guide the development of the law enforcement elements of
the plan. The FEMA, the Environmental Protection Agency, and the Department of Energy were to provide advice and assistance in the event of an
attack.

First responders were trained to provide rapid and targeted assistance to victims through their specialized equipment and pharmaceutical stockpile. Early in the training it was emphasized that a terrorist attack affects the entire community. A chemical attack may create not only victims at an attack site, but also an off-site plume of agent that could travel to residential areas. A nuclear attack would be regional with blast and fallout effects far beyond the bomb site. A biological attack would have community wide effects, with mainly psychological impacts for noncontagious diseases like anthrax, and medical community impacts with contagious diseases like plague and smallpox.

3.2.2. EVACUATION OR SHELTER IN PLACE

One of the tools that would be used to respond to an attack on the community would be population protection actions such as sheltering-in-place. In some cases people would need to seal themselves in their homes until the chemical or fallout plume blew away. In other cases people would need to stay at home to prevent the spread of disease. If there were a warning of an attack, evacuation of the target community following old civil defense structures might be the preferred method. Communities quickly realized that every choice carried with it drawbacks that were both practical and legal, but a threat of terrorism or an attack would require decisive action to save lives.

A crucial issue for the first responders of a nation under threat of terrorism is the issue of which population protection mechanism will be most effective: evacuating or sheltering in place. Each choice brings with it different dangers and different benefits. Emergency management structures provide for the authority to order evacuation or sheltering in place. In most nations the evacuation decision rests with the local government of the area for all natural hazards. However, it may be that if there is no MMST, or equivalent first responder organization, the local leadership may be unaware of the implications of a terrorist attack, and may not have the capability to choose effectively between evacuation and sheltering in place.

In some nations there is a constitutional mechanism that allows a higher level of government to order the evacuation to protect the residents. This power may be vested in a state or regional council or the federal government. The authority responsible for the response to the attack must have information available to justify either decision. One mechanism for determining the safest response for the residents is evaluation of intelligence by local law enforcement units in consultation with the federal law enforcement agency, such as the FBI. In addition two methods of assessing existing conditions are available: real-time air sampling using standard HAZMAT response equipment; and air quality analysis based on modeling, the technical aspects of which will be discussed later in this chapter.

Legal considerations also effect the decision to order an evacuation or sheltering in place. Ordering a mandatory evacuation in the USA requires that the law enforcement agency for that jurisdiction arrest people who do not comply with the order. Therefore in many California communities the government issues "evacuation advisories," notifying people of the need to leave, while not setting up an adversarial relationship with law enforcement. If evacuation is ordered, businesses will have to close, and they will suffer lost revenues. If these losses are sustained without a disaster occurring, there will be political backlash by residents.

Conversely, community leaders may determine that the risk of an event is not high enough to justify a mandatory action. They may issue watches or warnings, but not issue evacuation orders. Their caution may be based on the lack of scientific evidence, or strong enough probable cause to withstand a court test. However, if an event then occurs, the local officials will be blamed for inaction.

Sheltering in place is another option, which sends people into their homes until officials have evaluated the likelihood of an event causing high enough concentration of HAZMAT to harm humans. Instead of leaving the community, residents are encouraged to seal themselves into their homes with duck taped windows, covered vents, and sealed doors. One of the first acts of Secretary of Homeland Security Tom Ridge was to encourage Americans to make a shelter in place kits for their homes. Modeled on long standing advice to people who live downwind of refineries and other HAZMAT producers, the kit was to contain duct tape and plastics sheeting for sealing windows and doors against external pollutants, and a radio of obtaining up to date response information from the local authorities. However, misunderstanding of the advice led to panic buying of materials and misguided plans to seal homes (CNN, 2003).

Officials are also concerned that if people are ordered to shelter in place they may not comply. Either they may leave their homes instead, running into the polluted air, or they may not adequately understand the steps for securing their homes, and remain at risk. Another concern is how to respond to medical emergencies that could occur in homes where someone is sheltered, such as heart attacks, strokes or onset of childbirth. Officials could be blamed for leaving the patient in harm's way, as removing the person from the home into the polluted air could do serious harm.

If a threat is leveled against a community, sheltering in place would be the least disruptive mechanism for public safety, requiring the least activity by first responders, who could instead be focused on preventing the event if possible, and caring for the victims if it occurred. Evacuation would place the public in cars on crowded roads, possibly subjecting them to the greatest effects of blast, air pollution, and disruption. Also, people driving their personal vehicles during an evacuation would have a higher incidence of accidents due to the stress of driving in crowded conditions surrounded by people who are fearful, and "road rage" induced by fear of the oncoming event (*USA Today*, 2005). Vulnerable populations could be more severely effected, as when three nursing home evacuees died while being transported to Baton Rogue prior to Hurricane Katrina making landfall in New Orleans (Associated Press, 2005).

Liability for deaths and injuries resulting from mandatory evacuation would lie with the issuing authority. Political responsibility for damages due to either sheltering in place or evacuation would lie with the issuing authority. Failure to act could be considered dereliction of duty. Unnecessary evacuation or sheltering in place could be characterized as alarmist behavior. The crux of the problem is simply stated: who knew what and when did they know it?

Hurricane Katrina in the fall of 2005 provides a paradigm for evaluating the evacuation decision. The governors of Alabama and Mississippi joined the governor of Florida in an early declaration of state of emergency, and request for presidential disaster declaration on August 27. Mayor Ray Nagin of New Orleans issued a mandatory evacuation order the following day (Associated Press, 2005), as did other parishes. 80% of the population of New Orleans fled. Some people chose to stay behind, regardless of the warnings. "The head of Jefferson Parish, which includes major suburbs and juts all the way to the storm-vulnerable coast, said some residents who stayed would be fortunate to survive. 'I'm expecting that some people who are die-hards will die hard,' said parish council President Aaron Broussard" (Associated Press, 2005).

Yet people around the world have blamed local, state and federal officials for the deaths of those who refused to leave the drowned city. The House of Representatives' select committee issued its "Failure of Initiative" report in February 2006 criticizing the officials at all levels for failing to act early enough in evacuating the communities at risk (Hsu, 2006). The criticism of the evacuation being too slow demonstrates that almost any decision for evacuation or sheltering in place leaves the public officials at risk for criticism of their decisions.

Once the evacuation or shelter in place order has been issued, will the effected population know what to do? Community emergency plans are developed through the community Office of Emergency Services, civil dense office, or similar structure. Community emergency plans include population education through mechanisms such as CERT mentioned above. Pre-event education teaches people to keep their car gas tanks at least half full, develop a buddy system with able bodied friends or relatives if they have disabilities, and make an emergency kit with essential personal support items, medications, and financial documents that is ready to "grab and go" at all times. It also teaches them to make a shelter in place kit and plan for sealing their homes and businesses. Cooperation with preparedness advice will enable residents to evacuate or shelter in place during a disaster.

The government can create effective disaster response plans for all reasonably likely risks. Threat analysis is an integral part of emergency planning. Likely threats should result in response plans, training for first responders on appropriate protective and rescue actions, and regular exercises of the plans. Unfortunately, simply having paper plans and exercises does not guarantee readiness. For example, in 2004 FEMA sponsored a 5-day exercise in Baton Rogue, Louisiana called "Hurricane Pam." It postulated a Category 3 hurricane hitting New Orleans and causing overtopping of the levees (FEMA, 2004). Unfortunately, the lessons learned from the exercise had not been turned into action by the time the real Hurricane Katrina hit the community.

While the partners in the exercise agreed that the levees were a known weakness, the Army Corps of Engineers that is responsible for the levees did not have Congressional budget authority for the significant upgrades to the levee system that would be required to protect the city from a Category 5 storm. After Katrina the levee repair estimates are over $9 billion for basic repairs to allow residents to obtain flood insurance. A $20-million study is underway to determine the cost of a levee system strong enough to withstand a future Category 5 storm (Vartabedian, 2006).

3.2.3. FEDERAL STRUCTURES FOR RENDERING AIDE

In the USA there is a system for getting help from one level of government to another. Under the National Response Plan the federal government organizes its resources through "Emergency Support Functions" (ESF) (Department of Homeland Security, 2004). Each ESF has a lead federal agency and supporting federal agencies. For example, ESF #10 "Hazardous Materials Response," and is coordinated by the Environmental Protection Agency, with EPA and the Coast Guard as the Lead Agencies, and 13 other federal agencies as "Support Agencies." Among these are the Departments of Defense, Transportation and Nuclear Regulatory Commission.

Local governments whose resources are inadequate to the response task can ask for assistance within their state. After declaring a local emergency they request the governor to declare a state of emergency. If the governor determines that the state cannot meet all the disaster response needs, the governor asks the president for a presidential disaster declaration. The presidential disaster declaration opens the federal checkbook and the federal cupboards. The Stafford Act provides for the federal government to pay 75% of the emergency response costs and a substantial percentage of the repair costs for public infrastructure damage (Federal Emergency Management Agency, 2006a).

The local government first responders, acting within the local, state and national emergency management framework, respond to the scene of the disaster, using the resources available locally, and the training that they have received. When the disaster is terrorism they will rely on the locally based but federally funded Metropolitan Medical Strike Teams (MMST) to provide guidance on the protective equipment needed by the first responders, the organization of the care of victims from the scene through definitive medical care, and the protection of the community. There are also specialized federal teams to enhance local disaster response. For example, the Disaster Medical Assistance Teams (DMAT) (State of California, Emergency Medical Services Authority), include medical professionals and equipment that can augment local hospitals or replace hospital staff who need to care for their own families. Urban Search and Rescue Teams (USAR) bring construction, engineering, medical and fire professionals and equipment for specialized rescue tasks such as high angle rescue, collapse rescue and swift water rescue (Federal Emergency Management Agency, 2006b).

3.2.4. NON-GOVERNMENTAL ORGANIZATIONS' ROLE

Working together, these public safety professionals respond to the disaster to save the victims, protect property and help return the community to normal. They are augmented by the nongovernmental organization (NGO) of the community who provide human services to the victims. For example, the American Red Cross assists local governments in opening shelters for the displaced population. The Salvation Army assists with shelters for the vulnerable populations with special needs, such as the frail elderly. The NGOs have their own national organization in the USA that is focused on assisting them to become prepared to be part of the response to a disaster. The Voluntary Organizations Active in Disasters (VOAD) provides training and planning assistance to any non-profit organization that wants to become involved in disaster response (National Voluntary Organizations Active in Disaster).

In addition there are local organizations that bring together the voluntary community resources for disaster response. For example, in Santa Clara County, California there is the Collaborative Agencies Disaster Relief Effort (CADRE), a coordinating committee established immediately following the Loma Prieta Earthquake in 1989 (Volunteer Center of Silicon Valley). This group has divided up the disaster response tasks into ten "spokes," and has one lead agency in charge of each. Managing goods and property donations is the responsibility of the St. Vincent De Paul charity, while managing food donations is the role of the Second Harvest Food Bank.

CADRE provides its members with assistance in writing an agency emergency plan, and in training their staff on the Incident Command System (Federal Emergency Management Agency, no date) and the Standardized Emergency Management System (California Governor's Office of Emergency Services) used in California for all government response to disasters. They also hold regular planning and training update meetings, and tabletop exercises.

Vulnerable Populations

In 2003 CADRE joined with the local Emergency Managers Association to place a focus on planning for the special needs of vulnerable populations in disasters, since many of these agencies provide services to people with special needs every day. Issues such as mobility, accessibility and medical needs are addressed, along with language and cultural communication needs. They support the development of self-sufficiency for individuals with special needs, including developing kits of supplies at home, and arranging the furnishings for easy egress following a disaster.

In addition, California has licensure requirements for those who provide services to the special needs population. For example, companies that provide home medical support equipment, such as ventilators, must provide backup battery power supplies to their clients. Nursing homes are required to have an agreement with another nursing home outside of the immediate area for the relocation of residents during a disaster, such as a fire or earthquake. Finally, staff members receive training in managing residents during disasters, including the requirement for staff members to remain at work.

3.3. Using Technology to Aid Planning and Response

3.3.1. AIR MODELING[6]

Some hazards pose an inhalation hazard to the community. The ability to determine the locations of plumes of toxic material, their concentrations, and the movement of those plumes are all key factors in planning for the response to an airborne event. For example, the decision to shelter in place

---

[6]Material regarding the air modeling is based on a presentation by Stan Rydell delivered at the NATO STS-CNAD Workshop, Portugal, 2006.

versus evacuating the impacted population is driven by factors like the speed with which the plume will arrive in the neighborhood, the length of time it will remain at a high enough concentration to impact human health, and the ability to adequately seal a home against the material.

The fire service has had plume modeling capability in their HAZMAT teams for over 20 years. Systems like CAMEO and ALOHA[7] have been used on laptop computers for years to estimate plume deposition in a community. The original purposes were to estimate whether there would be off-site consequences during an industrial chemical accident and release, and to determine the definition of the "hot zone" where Level A personal protective equipment[8] has to be worn, and the "warm zone," and "cold zone," at an incident. Therefore the plume model assumed open conditions, and stable weather during the event, since it was assumed to be relatively short-lived. In addition the chemical was accurately identified in the required business emergency plan, and the maximum quantity that could be released was based on the storage capacity of the failed container.

In using plume modeling to help manage a terrorism event, the changes to the prediction parameters need to be considered. Most important, you may have no point of origin to use to orient the model. In addition, you may have only limited information about the chemical of concern. You may have air sampling capability that will define the class of chemicals, such as knowing that the item is an organophosphate, but you would have no information regarding its original concentration or form, and the amount that was released. Both of these factors will make the information from the model less reliable.

Furthermore, terrorist release of a hazardous chemical is likely to occur in an urban environment that is built up. This means that there are likely to be storm drains, streets with gutters, and tall buildings, each of which will impact the deposition of the material in ways that the traditional plume modeling software cannot account for.

The US Environmental Protection Agency (EPA) has studied the use of plume models in terrorism events. The US Department of Energy (DOE) has created additional models that better take into account the unique features of the urban environment, and the impacts on the deposition of a weaponized chemical there. These models consider the unique features of urban terrain as they effect the deposition of unknown chemicals. For example, storm drains and gutters may channel and retain liquid and vapor. Tall buildings

---

[7]The EPA website has a comprehensive discussion of the capabilities of CAMEO, ALOHA and related products. http://www.epa.gov/ceppo/cameo/.

[8]For a description of Level A and Level B personal protective equipment see http://www.sccfd.org/clothing_hazmat.html.

can cause "canyon" effects, creating their own microclimates including unique temperature bands and air circulation patterns that are external to the meteorology of the area. EPA and DOE staff members have created models for the urban environment that can be used in by emergency planners and first responders for both the planning phase and the response phase of an urban terrorist event involving a chemical weapon.

During the preemergency phase, when plans are being developed and resources coordinated, modeling can be used to provide a variety of facts useful in developing the best response. For example, models are used to determine the likely people and areas to be impacted. They can predict the time factors of impact on people and places. The model values can then be charted for comparison with existing emergency plans. Areas with mismatches or gaps can then be examined by the responders, and a better plan can be developed.

The model factors that are most useful are mainly the human health and safety factors, such as those that influence release characteristics that impact urban response planning. For example, models can predict the distribution of causative substances and the probable health and safety outcomes from human interaction with these substances. The models can then predict how to select among the shelter in place or evacuation choices available to the Incident Commander or the Emergency Operations Center staff members. Predictive models can be used to give estimates of the scope of losses, and damage, including an estimate of the population impacted, for testing the adequacy of emergency preparations.

Contemporary terrorists are often willing to trade their lives for the success of their attack, as demonstrated by suicide bombers in Israel and the 9/11 attackers in the USA In the past emergency planners assumed that terrorists would want to get away with their attack, which ruled our materials that would contaminate and kill them, such as radiological and nuclear materials. The current radical Islamist terrorist groups have no such limitations on their choice of materials. Emergency response planners have to be prepared for this new reality with better estimates of the possible deposition of all the probable weapons of mass destruction/disruption/ killing materials (WMD).

The use of the models has several desirable outcomes for the emergency response planners. First, modeling allows planners to make educated estimates of which assets are most vulnerable to terrorist WMD attacks. Second, this leads to an analysis of what assets to defend, either through pre-event installation of protective devices, or through planning for rapid mobilization of post-event assets. Third, knowing the way that the WMDs are likely to be deposited in the urban environment after release will inform decisions on how to defend human and other critical assets. This will then lead to an evaluation of what people and equipment are needed to counter the use of the WMD. It will also assist planners in determining what emergency exercises to run. The model can test scenarios to ensure that they are "real world" examples of likely events. Finally, this chain of activities will lead to a heightened state of emergency preparedness for WMD response.

Four models offer the best results for predicting and planning for response to a WMD in an urban environment. They are Hazard Prediction and Assessment Capability (HPAC), Catastrophe Assessment Tool Set (CATS), Nuclear Biologic Chemical Casualty and Resource Estimation Support Tool (NBC-CREST), and Oak Ridge Evacuation Modeling System (OREMS). These models are appropriate for use by all NATO members. The models each have worldwide map coverage, and they are available to NATO members (licensed). The models themselves are free, and the supporting geographic information system (GIS) costs about €2,000–5,000. The models run on a live weather download that is available daily around the clock from free meteorology servers.

Because each model is a bit different, different ones work better for different applications. For example, HPAC modeling might be preferred for chemical or biological weapons, chemical or biological facility accident/release, or a radiological weapon, and works well for planning and prediction. CATS is useful for a broader range of hazards, including earthquakes, hurricanes, or chemical, biological, radiological, nuclear or explosive (CBRNE) events, but is not predictive. This model creates and manages spatial data through ESRI ArcGIS 9. It assists with developing responses to evacuations, isolation of hazards, roadblock listings and casualty reports. CATS can exchange data with HPAC. CATS provides a graphic depiction of a probable plume direction that estimates the area where death is possible due to the lethal concentration, the area where injury is likely due to the high concentration, and the plume that extends downwind with adequate concentration to reach Threshold Limit Values for exposures. It models the expected population in the contour and the length of time that the hazard would continue. This helps to determine the response plan. It also allows the data to become specific for daytime and nighttime populations. This is a critical planning factor for major cities, with large areas of commercial activity during the day that are mostly empty at night. The CATS data is more accurate, and therefore allows for the planned concentration of response assets in the area where harm is most likely.

When compared with ALOHA and CAMEO, the CATS plume model is more focused. Using ALOHA or CAMEO would involve a larger population base that is not likely to be at risk, wasting time and resources on people who would not have been victims, and dislocating people unnecessarily. This can become a life safety issue when people are moved unnecessarily from nursing homes, hospital intensive care centers, and senior assisted living facilities. The stress of the move may be fatal to the frail elderly or people on ventilators and medical gasses. Therefore, limiting the population evacuated to only those directly at risk can both save lives and focus resources more appropriately to those in greatest need.

OREMS provides a tool for estimating the size of the population that needs to be moved and the routes that would be best for the population movement. It takes into account the time of day, the population of the impact area, the likely number of people per car (based on population data), the number of vehicles per hour that a road can carry, and weather factors. It can also make assumption about the time it will take to make the evacuation decision after the attack has occurred, and the time it will take to mobilize for the evacuation. For example, changeable sign board messages, traffic signal reprogramming and placement of tow trucks would all have to be accomplished before the population was asked to start an evacuation. In addition, reception sites in outlying communities would have to be opened in time to receive the fleeing population. The model factors in links and nodes in the road system to create the best routes to move the most people out of harms way in the least time.

Model values are a good check on population and time values used in emergency preparedness plans. Application of GIS and meteorological data make for modeling output with a high statistical confidence level. HPAC, CATS and other models produce good results, and there are experienced users or expert advisers available 24/7. However, model users need to understand that mixing models and types of data can lead to low confidence levels and ambiguous results. Planners should also realize that because of the lack of specificity, forecasting outcomes for possible future incidents using various time and related population parameters can be more work than modeling for the same event in real-time, when actual parameters are available.

3.3.2. LAND USE-BASED MODELING[9]

Land use in a community has a direct impact on planning for disaster response of all kinds. High and low population densities, location of critical infrastructure, and the existence of WMD components in the community are

---

[9]Material regarding the GIS applications includes information from a presentation by James David Ballard delivered at the NATO TS-CNAD Workshop, Portugal, 2006.

all important factors in predicting mechanisms of disaster, likelihood of disaster, and populations at risk.

Since the 1980s American cities have been using computer-based inventories to maintain records of their critical infrastructure, roadways, parks and public buildings. These systems allow maps to be created that include many separate layers that can be integrated to create visual depictions of the community in various ways. For example, a map can be made that shows the street grid, the earthquake fault lines, areas prone to flooding, wildland urban interface fire threat areas, HAZMAT storage facilities, schools, hospitals, and fire stations. These maps are used by emergency planners as they anticipate the emergency response that will be needed after a disaster, and by the fire department to chart which roads are most likely to be passable. Wall maps showing these community characteristics have long been a staple of emergency operations center (EOC) resources.

Maps can also show multiple layers, such as the streets, the underground utility lines, and the water and sewer lines in the right of way. Such maps assist emergency responders in anticipating earthquake damage to these services.

Geographical information system (GIS) maps have added the capability to create customized maps in the EOC during a disaster event. GIS technicians can build up a map—using layers of previously stored data—as a disaster unfolds, and then use an LED projector to display the map on a large screen visible to all the staff in the EOC. The customized map may also be sent by e-mail, printed out, or faxed to another location.

As the GIS capabilities have become more sophisticated, software tools have allowed for the integration of GIS into other operational systems, such as the computer-aided dispatching (CAD) systems used by police and fire personnel. While a fire company is on the way to a response they can access the utility layers of a map to anticipate the location of the closest hydrants and the location of gas utility shutoff points for a neighborhood. A police department special weapons and tactics team (SWAT) might use GIS maps accessed through the CAD system to estimate the proximity of an active criminal to high population buildings or schools, or the addresses of homes within range of the criminal's weapons to determine who should be evacuated.

As terrorism has become another hazard to be planned for, GIS maps are useful for understanding the relationships among transportation routes, utility hubs, and critical infrastructure. Multilayer GIS maps enable emergency planners to design realistic tabletop exercises for first responders, incorporating the realistic consequences of an attack on a specific location. Examination of the maps also leads to a clearer understanding of the interrelated vulnerabilities within a community. GIS maps provide the opportunity to merge topography, land use information, and critical facility locations in real time to assist with planning the response to an attack on a likely target.

For example, the City of San José, California, held an exercise of its response to a terrorist attack on a well-known public building. The Incident Commander asked for maps that showed the street grid around the building for several blocks, and the critical utility connections, high occupancy buildings, and location of vulnerable populations in the area. This allowed him to determine which buildings had to be evacuated, and what direction the people could go to avoid being exposed to the terror attack's off-site consequences. In addition, the maps showed which utility connections might be at risk from the building where the event occurred, and these were either shut off or protected.

For pre-event planning, maps can be used to estimate population for the placement of disaster response supplies, and the designation of shelters. GIS maps may be created to show the zoning of the community, which displays the land use of the area, the density of that land use, and the relationships of various uses to each other. They can help to estimate synergies among critical infrastructures that enhance their vulnerabilities, such as HAZMAT facilities that are close to major highways, leading to plans for more effective mitigation and protection.

Layers of information can be obtained from other government sources. For example, the San José Police Department CAD system has access to the County Assessor maps. These show the use and occupancy of a building, its assessed valuation, and the owner. The Santa Clara County Sheriff's Department has GIS maps of all the public high schools, showing floor plans and utility connections, which can also be accessed through the CAD system.

New technologies such as satellite imagery are also being integrated. While aerial photographs used to provide information on roads, vegetation and facilities, they were a static record that was quickly outdated in a growing community. The Google Earth satellite mapping system (www.earth. googlecom, 2007) provides up-to-date satellite images that show current conditions in great detail, including maps, terrain and buildings in three dimensions (3-D). This capability was accessed during Hurricane Katrina's response phase to allow people in the diaspora across the country to see whether their neighborhood still had standing water, whether their home was still standing, and whether there was access available to their home.

Because of the rich resource of information that GIS maps represent, many cities have taken their maps off the World Wide Web after 9/11. For example, San José used to have a map showing all its water system connections on its Public Works Department website. This information was useful to contractors who were building or remodeling buildings, because it showed which water company line served the area where they would build, and where they could tie in to existing utility lines. After 9/11, these maps were removed to prevent a terrorist from using the map information to access the water lines for the purpose of introducing a toxic material into the drinking water supply.

GIS capability is a new resource for disaster response planning that allows for the integration of information from a variety of sources for planning, preparing and responding to any disaster, including WMD terrorism.

3.3.3. STRUCTURAL ENGINEERING AS MITIGATION

Another application of technology that is an important part of terrorism preparedness is structural engineering. Building codes have been developed to ensure that structures are safe for occupancy. The codes take into account the specific conditions within a community, such as soils, topography, and physical features that may cause disasters. Cities near wildland urban inter-face zones have stronger regulations of building and roofing materials. Those in flood prone areas will require higher elevation of the first floor of living space. Communities in proximity to earthquake faults will require designs that are more flexible to withstand earthquake-related shaking.

Because escape from a multistory office building during a fire would be difficult, the fire code in most communities requires that high-rise buildings be built of fire resistant materials. Until 9/11, no high-rise building in the USA had ever collapsed due to a fire. High-rises are built with fire resistant steel construction, and normally have sprinkler systems to quickly suppress the spread of fire. Staged evacuation was the normal procedure, because it was assumed that the building would withstand a fire burning on only a few floors. Following the traditional procedures after the airplane crash on 9/11 led to significant loss of life as the towers collapsed.

The failure of the World Trade Center towers following the 9/11 attacks was due in part to the structural engineering of the building. Although the building had been designed to withstand a direct hit by a full loaded 707 airplane, the planes that hit the tower were a newer and larger model with a greater fuel capacity (Kennedy and Klein, 2002). The fireball of jet fuel that poured through the building's elevator shaft resulted in a rapid spread of fire that took the temperature within the building to over 1700 degrees Fahrenheit. The force of the crash knocked much of the asbestos fire proofing off the steel girders, making them more vulnerable to the intense heat. As a result the steel stretched and the towers fell (World Trade Center Building Code Task Force).

The National Institute of Standards and Technology (NIST) conducted a 3-year study of the collapse of the World Trade Center towers. This effort included interviews with over 1,200 people, computer simulations of the buildings' performance following the attack, and a review of documents related to the buildings and their construction (NIST, 2005).

The NIST study resulted in 30 recommendations for future building construction, with the goal of increasing structural integrity. Some recommendations are for the enhanced fire endurance of buildings. Others cover new methods for fire resistant design, and enhanced active fire protection. Emergency management practices were also evaluated and improvements noted.

The significance of the NIST report is that it is an acknowledgement that building design and construction should be considered as a mitigation method against disaster. Structural engineers in seismically active parts of the worlds have long planned for the inevitable earthquakes by strengthening building designs to withstand the anticipated worst-case earthquake. Structural engineers in hurricane prone parts of the world likewise design their buildings to be wind resistant. Since there is no "terrorism prone" part of the world, structural engineers in all nations should consider how buildings might become more resistant to the types of weapons used by terrorists.

Building design in high hazard areas points the way for structural mitigation that can be applied to resistance to damage from CBRNE materials. Thick concrete walls resist radiation. Buildings that are built to be able to be sealed off from outside pollutants, including toxic industrial chemicals released during accidents, provide a design concept that can be applied to all buildings in high risk areas for terrorism. Air handling systems can be built with sensors for chemicals and biological materials, and located on rooftops or in other inaccessible areas.

Similarly, best practices for emergency response can be incorporated into building construction. Radio antennas to ensure that first responders can communicate within the building are a relatively inexpensive addition during the construction phase. Air lines for refilling fire fighter air bottles can be installed with the building's plumbing system, with outlets in each fire stairway on each floor. Polyester shatter resistant film, such as 3M Scotchshield, can be placed over safety glass to prevent damage to personnel and building contents from flying glass after an explosion (Brower Window Films, 2007).

Chapter 6 provides additional ideas for innovation in the design and construction of the built environment to better protect the population from the depredations of CBRNE terrorism. Engineers, building designers and materials engineers should join together to consider ways that the built environment can truly become shelter from terrorism in the urban centers of the world. NATO partners working together have the opportunity to create safer workplaces, schools and public spaces as future development occurs.

3.4. Cross-Border Interoperability Across Systems

NATO member states are prepared to deliver assistance to each other during disasters, including terrorism. However, providing assistance involves crossing international borders, which raises a variety of international and cross-border issues. Although many of these will be raised in more detail in Chapter 4, a brief summary here serves to highlight the challenges facing the international community as it rallies to aid the victim nation. For example, each sovereign nation makes its own rules about the weapons that may be introduced into it. Under NATO and other treaties these rules are suspended during war and for military units responding to an attack by a foreign power. Does a response to a terrorist attack also provide for a suspension of border crossing regulations for armed troops? Do these suspensions of customs rules also cover law enforcement personnel responding to a terrorist attack?

International terrorism is a hybrid of war and crime, with rules of engagement and response that are still to be fully codified. Until the decisions about weapons and equipment that may be imported are sorted out. Cross-border assistance becomes a challenge for emergency management. Existing emergency management structures were designed for mutual aid within the nation, provided by first responders with the same training and similar equipment. Cross-border assistance poses challenges for both law enforcement and medical caregivers. Professional licenses, weapons, drugs, treatment modes, even those who can treat patients, are all impacted by the laws of different nations or states. If a mutual aid structure is to be developed for terrorism, similar to the treaty-based aid that is available for military attack, then a formal mutual aid system and agreement must be developed soon to facilitate the transfer of technology, personnel and logistics support from donor to victim.

Some systems are clearly more mobile and raise less law enforcement, customs and licensure issues. However, they still pose problems for interoperability among responders from different nation-states.

3.4.1. COMMUNICATIONS SYSTEMS

First responders typically rely on radio using frequencies assigned or controlled by the national government. In the USA, for example, police and fire departments have historically been on different frequency bands to prevent overlap in radio transmissions during day-to-day responses in the community. This lack of a common frequency has cause significant problems when there is the need for a joint command structure and coordinated action across disciplines.

For example, following the attack on the World Trade Center, the Fire Department of New York, the New York City Police Department and the Port Authority of New York and New Jersey Police were using radios on different frequencies. Therefore, members of the other departments could not hear messages being passed, resulting in limited situational awareness. The most tragic example of the danger this poses is the fact that firefighters inside the World Trade Center towers could not hear the police officers in the helicopter questioning the safety of the tower. They continued to enter the structure and try to go to higher floors, while police officers were evacuating the buildings. This is one reason why so many firefighters lost their lives as the towers collapsed (O'Hare, 2002).

A similar problem will occur when first responders from different nations respond to the same event. There is a need to create a bridge between the different frequencies to ensure that messages are heard by all professionals at the scene. In San José, California, the police and fire departments within the county have developed a system using shared frequencies and microwave dishes to enhance radio signals. Upon arrival at the scene of a disaster a set of common frequencies is assigned for use by all agencies during that event (BAYMACS, 2003). This system, known as BAYMACS, enables all responders to communicate across jurisdictions and professions. However, on an international basis the development of a frequency sharing system may not be practical, since the mutual aid partner agencies may not be known in advance of the event.

The US government has begun an aggressive development program for more effective interoperable voice communications. Known as SAFECOM, its goal is to create new technologies and methods for enhancing communications among agencies at a disaster. They have outlined an "Interoperability Continuum" as a pathway for working toward this goal (Department of Homeland Security, 2004). The continuum acknowledges the importance of starting with existing technologies because of the cost of developing a completely new system. Steps include the simple development of a cache of radios that can be swapped during a response. Gateway devices, shared channels and proprietary shared systems are incremental steps. The final goal is standards-based shared systems that would be in daily use throughout a defined region to ensure familiarity of the system to all first responders.

Having a cache of spare radios for swapping is the least costly approach to interoperability. However, it is a challenge to keep all the radio batteries charged when they may only be used once or twice a year. Furthermore, using an unfamiliar radio set during a disaster may add stress to the first responders and lessen their efficiency and effectiveness at the scene. One way to overcome the issue of unfamiliarity is to use a computer-based gateway, which coordinates the frequencies used by the various agencies at the scene (Reyes, 2006). This technology is effective for a single site event, but may not provide an acceptable level of service for multisite disaster response.

Another option is the use of existing WiFi systems with laptop computers to exchange data over the Internet. San Mateo, California police department has had a WiFi-based system with laptops and aircards running since 2003 (Swider, 2003). New Orleans and Corpus Christi also have this system. While not permitting effective voice communication, this WiFi "MetroMesh" does provide a good solution for the exchange of information in data formats (Tropos), such as photos, missing persons reports and detailed criminal records. Third, some communities, such as Anaheim, California, are using portable repeaters known as "bread crumbs" to establish a proprietary network in the field at the scene of a disaster. The "bread crumbs" can be set on car roofs or other elevated locations to carry the signal further from the scene. This system is the most flexible, and can be operated over large areas using multiple repeaters (Wills, 2007). This system will support voice communication, as well as video-streaming, text messages and e-mail.

All of the communications technologies share common challenges that provide areas for future research and improvement. First, with the development of so many radio frequency-based electronic consumer technologies, most nations are experiencing frequency congestion. Finding available frequencies that are appropriate for secure communications is a challenge. There may need to be international protocols for the use and sale of radio spectrum to ensure that first responders can be adequately supported to make the best use of all technologies available to enhance their life saving activities.

Second, there is the problem of radio discipline in the field. Each agency has its own protocols and standards for managing radio traffic. Some use dispatch centers to manage the voice traffic. Others have chain of command-based protocols. When multiple agencies gather to manage the same disaster they will need to determine what protocols to use for message traffic, and how to differentiate and prioritize among emergency messages, logistics messages and other types of communication.

Third, cellular phone systems were once considered a good alternative to public safety radios for interoperability. However, in recent exercises and in real events it has become clear that they are too unreliable for life safety activities, and that they stovepipe information completely. The unreliability of cellular phone systems has been proven at the World Trade Center attack on September 11, 2001, during the Hurricane Katrina response, and in California's wildland interface fires. The system gets overcrowded with calls made by the public. Systems often fail because of loss of power or loss of cell sites. For example, on September 11, 2001 the loss of the antenna on the Verizon Building on Chambers Street and the antennas atop the World Trade Center caused the New York metropolitan area to lose cellular service for hours (Dawes *et al.*, 2004). During Hurricane Katrina most cellular systems failed within the first few hours. The cellular system lost power throughout the disaster area, and the backup generators ran out of fuel or were flooded. Furthermore, many of the cell sites were installed on unstable towers and fell over during the hurricane and flooding. Days after the disaster the cellular service was still non-functional (ConsumerAffairs.com, 2005).

The problem of information "stove piping" has been recognized through a state-level multiagency exercise. During the exercise some of the first responders used cellular phones for exercise communications in order to lessen the message traffic on the normal emergency frequencies. The result was that the level of situational awareness among the drill participants fell dramatically. First responders are accustomed to monitoring their radios for all message traffic, even that not directed at them. This "party-line" arrangement allows them to keep up on developing response and safety issues without having to initiate or receive calls, which saves time and duplication of effort. When the leadership of the drill used cell phones the participants lost the information trail, and found that they were unable to make appropriate adjustments to their plans. In the "hot wash" meeting after the exercise the participants agreed that cellular phones would be excluded from all future exercises and actual responses to prevent this limitation of information (Edwards and Goodrich, 2006).

3.4.2. INTEROPERABLE EQUIPMENT

When agencies work together at the scene of a disaster they rely on a catalog of equipment that will differ from agency to agency, and from nation to nation. While crews can work side by side with their own equipment, it is more effective to have common equipment. Developing some standards for interoperable equipment would enhance disaster response, especially in cases of WMD attacks.

Victims of chemical, radiological and some types of biological events could benefit from being externally decontaminated, to ensure that they are not further sickened, and that they do not carry material away from the scene. Decontamination equipment varies widely among agencies and nations. Some use plain water while others use soap or a neutralizing solution. Some capture the runoff water while others do not. Some can provide heated water for critically injured patients with unstable body temperature, and first responders who are in the water for long periods of time. These different systems have differing power requirements, different water pressure requirements, and different setups.

Another type of equipment that would benefit from interoperability is personal protective equipment. Some countries have standards for various levels of protection from chemical contamination, with Level A being the highest. Nations that use this standard have training requirements for those using the encapsulating suits, and fit testing for the respiratory protection equipment. Self-contained breathing apparatus using supplied air is part of the ensemble, and the air bottles, masks and fittings are unique to each manufacturer. Air carts are required to refill the bottles. Some nations use military "Mission Oriented Protective Posture" (MOPP) gear, which is not designed to provide a high level of personal protection, but is designed to be lightweight and permit greater mobility. These rubberized suits include only respirators, in which canisters filter the ambient air. First responders working side by side at a WMD response would have significant differences in protection being offered by their suits. In a nation that requires Level A protection for unknown chemicals and high concentrations of chemicals, the responders with MOPP gear would have to work out of the hot zone. This kind of exclusion could lead to management problems on the disaster scene.

Adopting some common standards for critical response equipment would make cross-border assistance more efficient at the time of disaster. For example, Italy has developed caches of equipment in each national region that are all the same, so that when the different agencies meet for a large-scale disaster their equipment is all identical, and therefore interoperable (Steinhäusler and Edwards, 2006). The Italian equipment is interoperable with the American equipment, as well, which would enhance coordination at a multinational response.

Another element of interoperability is resource typing. Each piece of equipment for WMD response could be named and typed according to its response capability. For example, in the USA and in Italy, a "Level A suit" has the same characteristics in any place where the terminology is used. It provides for both splash and respiratory protection, has an undergarment of Tyvec and an overgarment of a chemical-resistant material. The suit accommodates the breathing apparatus and spectacle holders for prescription lenses. It has boots and gloves, and may require a cooling vest in hot seasons or climates.

The USA has had resource typing in the fire service for over 20 years. This allows an Incident Commander at a mutual aid event to request exactly the resources that he needs, and be assured of getting exactly what he asked for. The FIRESCOPE organization (FIRESCOPE, no date) has been responsible for overseeing the development and maintenance of resource typing. As part of the National Incident Management System mandated by

Homeland Security Presidential Directive-8 (Bush, 2003), all American first responders must participate in resource typing across disciplines.

Resource typing includes 120 resource definitions based on category, kind and type for each item (Federal Emergency Management Agency, 2005). These definitions, kinds and types must be used when making mutual aid requests through the national response system. For example, Type 1 Hazardous Materials Entry Teams are mandated for all WMD events (Federal Emergency Management Agency, 2005). A table on page 15 of the resource typing document lists the "Equipment/Protective Clothing: Ensembles," based on National Fire Protection Association Standard 1991. This standard is readily understood in every state and community with a HAZMAT response capability. When an Incident Commander requests a Type 1 Hazardous Materials Entry Team, he knows exactly what he will get.

3.4.3. COORDINATION/COMMAND AND CONTROL SYSTEMS

Civilian first responders are being confronted by the terrorists organized outside of the nation-state model. Their approach, with its roots in Mao Tsetung, and an approach that rejects the conventions of a nation-state, is characterized as fourth-generation warfare, which creates new challenges for first responders. Generational warfare principles are discussed at length in Appendix 2. Combating this swift, singular, civilian-oriented brand of warfare puts civilian first responders on the front line, with challenges that will force an evolution of command and control toward a more military format, such as the Incident Command System currently used in Italy and the USA.

Terrorists strike with little or no warning. Their weapons are covert and often suicidal, aimed at rendering maximum damage on a defenseless population. Their operational tempo is focused on the first few minutes to deprive the victims of much chance of escaping injury and death. This means that the first responders have no warning of an impending disaster. They do not know until they arrive and review the scene that the event is an attack, and not an accident.

For example, the first responders to the World Trade Center on September 11, 2001 assumed that the first plane that hit the North Tower was an accident. For the first 17 minutes, until the second plane hit the South Tower, the explosion and fire was treated as a large fire. Over 1,000 firefighters were dispatched to the scene for a rescue mission only, and evacuation of both towers had begun. With little information about conditions on the fire floors, but awareness of the damage to the elevators and windows, the command determined that it was not possible to fight the fire (National Commission on Terrorist Attacks upon the United States).

In 17 minutes lives were saved and lost, and decisions were made that changed the course of thousands of lives. The fire command staff had to bring organization out of chaos without waiting for the normal command support structures. They had to confront a fourth-generation terrorist act with a mixture of first and second-generation warfare techniques, and some third-generation capabilities based on their specialized training. While the fire department used the Incident Command System, it was not interoperable with the city or port authority police departments.

Scene management also posed a unique challenge. The fourth-generation attackers chose their targets, weapons and timing without having to receive approval from an organized nation. The brutality of their attacks challenges the first responders to anticipate their methods and be prepared for secondary devices, such as the second plane. In fourth-generation warfare the secondary device is aimed at debilitating and killing as many first responders as possible, as a means of further terrorizing the victims. This forces the first responders to divide their resources into a response element, a group held in reserve in case the first responders are killed by a secondary device, and a group dedicated to seeking and disabling any secondary devices. Forcing a three-part allocation of scarce responders further slows the rescue and care of the victims.

The Incident Command System is an organizational structure with a flexible command and control element that can use first- through third-generation warfare functions. ICS also allows for the incorporation of different generational warfare concepts into the response. The Incident Action Plan lays out the goals for the next Incident Action Period. This empowers the various members of the Incident Command structure to develop their section, unit or branch plans based on their individual organizational needs to achieve a goal, rather than to simply follow a detailed order.

Further, ICS is based on a checklist approach to management. The Field Operations Guide (Incident Command System, 2004) provides checklists for each position in the ICS structure. However, the checklist is more like a series of questions than a series of orders. The ICS staff member can evaluate the checklist item based on the situational awareness available to him at that time, and make response plans for that Incident Action Period. To the degree that response actions at a scene of terrorism can be initially managed through third-generation systems, the first responders have a chance to overcome the chaos of fourth-generation warfare.

3.4.4. RECEIVING EXTERNAL HELP—EXISTING SYSTEMS

There are already significant cross-border mutual aid systems in place. NATO has an emergency response organization called Euro-Atlantic Disaster

Response Coordination Centre (EADRCC), the United Nations has the Office for the Coordination of Humanitarian Affairs, and the World Health Organization (WHO) and the European Union (EU) have similar agencies. All of these are discussed in Chapter 4. These organizations have focused on humanitarian efforts on behalf of the victims of natural disasters. However, the challenge of trans-national terrorism has required these organizations to consider their roles in the care of victims of terrorists, as well. While some of the organizations have refused to plan for involvement in WMD events, others are taking a holistic approach to disaster response.

In many cases the multinational organizations are staffed by volunteers. Will a response to a terrorist attack effect the volunteer response? Organizations must ensure that volunteers are trained in self-protection, terrorism awareness, and the special needs of victims of WMD incidents, to mitigate the shock effect of an attack. Organizations run with paid staff may be delivering this information as part of ordinary training for their workers. If not, this information would also have to be provided to them.

Language barriers are also a problem in mutual aid and cross-border responses. First, there are likely to be language barriers within the first responder community as mutual aid resources come together. Plans made in advance should specify how language barriers will be overcome. Having a common command structure, such as ICS, will ease language issues by allowing first responders to work off Field Operations Guides in their own language but with the same page numbers.

There may also be language barriers between the first responders and the victims and residents of the area that was attacked. Plans should include the involvement of local first responders as translators and the principal interface with the victims. People who have been traumatized and badly hurt are unlikely to be able to manage a multilingual environment of care. Local bilingual volunteers organized for other social and humanitarian purposes should be included in the resource plans for WMD events.

In addition, nations and areas with a large tourist presence must plan separately for caring for people without local health insurance, with a language barrier, and possibly without a person to sign consent for treatment forms. Unaccompanied minor children of dead tourists will be a special concern, as demonstrated in the South Asia tsunami of 2004, which generated 39 unaccompanied minors, some tourists (Nurbaiti, 2005).

There are also cultural barriers to consider in cross-border terrorism response. Plans should be made ahead of time to orient first responders to the cultural norms and sensitivities of victim nations while they are en route. Gender role norms are a special area of concern, as Western first responders will include both genders, while receiving cultures may have barriers between unrelated men and women, or limits on what unrelated women can provide to men. For example, in some cultures women are secluded from unrelated men, while in others women may not touch certain areas of a man's body, such as the top of his head. In other cultures crossing your legs to show the bottom of your shoe can be offensive. Also the roles of men, women, and children may dictate the order in which non-emergency patients are to be treated. Such cultural issues can be addressed in Field Operations Guides created in advance of need.

For example, the Italians have a plan to receive external help. They have adopted the Incident Command System, which includes a liaison officer whose job is to coordinate with outside agencies, both internal mutual aid and external international assistance. Their operations plan for mutual aid also includes provision of translators for the Incident Command structure. Finally, through the Mutual Aid Coordination System (MACS) there is a method for coordinating and managing the goods and materiel provided by outside organizations. MACS allows for the reception and tracking of personnel, equipment, and goods to ensure the timely return or replacement of non-expendable items (Steinhäusler and Edwards, 2005).

However, in order to make the systems most effective the international community needs to determine what types of events, or level of event, triggers a request for outside aide. NATO has structures in place through EADRCC to respond to WMD events. While advance coordination may be successful for WMD events that are confirmed, are there other circumstances, such as preemptive action, that might qualify for cross-border mutual aid?

Since the costs "lie where they fall" according to SHAPE in Chapter 4, there may be a need to reevaluate how extended responses and expensive equipment are reimbursed. For example, in the USA, mutual aid is provided without charge for the first shift. After that time the receiving jurisdiction must provide for the feeding and housing of the first responders, and for any overtime premium above normal salary. They also replace any equipment that is consumed or destroyed. This financial arrangement helps smaller agencies to afford to donate the use of their capital goods and their personnel, since they are assured that they will not sustain major financial losses.

3.4.5. SPECIAL CHALLENGES FOR CROSS-BORDER LAW ENFORCEMENT ASSISTANCE

Law enforcement agencies have special concerns about cross-border operations. Earlier the issue of bringing weapons across borders was addressed. Different nations have different rules about the use of automatic and semi-automatic weapons by civilian law enforcement personnel. There are also different rules about the kind of training that an officer must have to use certain weapons. Obtaining adequate supplies of appropriate ammunition may also pose a challenge.

Other operational considerations must also be resolved before cross-border law enforcement mutual aid can be undertaken. Nations have different standards of training for their law enforcement officers. Each country has different constitutional standards for probable cause, due process and admissible evidence. Treatment of prisoners, suspects and witnesses differs widely among the NATO alliance. Finally, standards for cross-border pursuits and apprehension of suspects not on their soil is another challenge to international cooperation.

Some first responder agencies will not want to go into a terrorist attack zone without their own law enforcement component that is integrated with their training and operational plans. In the USA the MMSTs contain their own law enforcement branch that travels with them. This branch provides protection for the first responders en route to the terrorism scene. They provide force protection, scene control and evidence protection duties as the response is mounted. Most MMSTs will not deploy outside of their jurisdiction without their law enforcement branch.

In an effort to address these cross-border law enforcement concerns Interpol is developing a response guide. In July 2006 they held their third workshop for international law enforcement agencies to discuss issues of common concern. The workshop and guide address law enforcement activities related to WMD terrorism responses, and will become a resource for law enforcement agencies working as part of international cross-border responses to terrorism (Interpol, 2006). A more detailed discussion of law enforcement issues is included in Chapter 2.

4. INTERNATIONAL DISASTER RESPONSE

NATO member states share two common experiences that have resulted in similar approaches to all hazards disaster response, and especially managing the response to terrorism. First, they all share in the military heritage derived from the Roman Legions and the military organization of Napoleon. As a result all acknowledge the importance of having a strong command and control function with highly differentiated functional sections. Second, the experience of World War II has demonstrated the importance of having immediate response capabilities within each local jurisdiction, including civilian involvement in the response, while also relying on the central government for the most specialized equipment and resources.

Terrorism against the NATO member states has escalated from single actor attacks on public facilities and personages, exemplified by the Irish Republican Army's long reign of terror against British facilities and personnel, to multiple armed attacks simultaneously, as in London's 2005 subway and bus bombings. The NATO member states have developed a highly sophisticated military capability that includes interwoven transnational forces, such as those being used in Afghanistan. This sophistication of military might deters opponents from attacking any NATO member state directly in a classic military confrontation. Instead, the guerrilla-based concepts of fourth-generation warfare[1] have become the guiding principles upon which NATO's adversaries are basing their attacks. First world nations, with their economic might and social stability, make poor targets for direct military action, but are highly susceptible to small bands of violent actors willing to die for their cause.

Recent catastrophic natural events have demonstrated that, for preparedness for terrorism to be most effective, it must incorporate capabilities for response to all hazards. The 2004 tsunami in southern Asia required the international community to contribute its search and rescue, medical and recovery capabilities to a whole region. Many NATO member states played key roles in providing specialized expertise, such as body recovery and identification (Scanlon, 2006). Later that same year, Hurricane Katrina in the southeast USA resulted in NATO's first combined effort in response to a natural disaster, followed closely by the NATO response to the earthquake in Pakistan (Bretschneider, 2006). The rapid deployment of trained and equipped personnel demonstrated the ability to be highly mobile, fast, and

---

[1]See Appendix 2 for a complete description of fourth-generation warfare, and the "generation warfare" theory.

F. L. Edwards and F. Steinhäusler (eds.), NATO and Terrorism, 83–159.
© 2007 Springer.

technically capable in the field with no notice of the event. Such responses offer a paradigm for the type of immediate local, regional, national, and international response that will be required by a catastrophic terrorist attack with weapons of mass destruction, weapons of mass disruption and weapons of mass killing, whether nuclear, radiological, chemical, biological, or explosive.

4.1. United Kingdom[2]

An overview of UK national emergency preparedness identifies strengths and deficiencies in national counterterrorism organization, law enforcement, fire services, local government, civilian management, and emergency medical services. The case study of the London July 2005 bombings presents challenges for national preparedness and for first responders. Future scenarios are outlined to describe a dynamic battle space that confronts first responders within the UK who are dealing with a global terrorist capability that has no national boundaries. Particular risks are identified from aggressive terrorist media psychological operations that use spectacular attacks for recruitment of suicide bombers and financing of terror plans. The conclusions address the challenge of embedded home grown terror networks, plans, and technology for first responders and the potential for overwhelming terrorism by further energizing public capability, building international real-time information networks and delivering cooperative campaigns.

4.1.1. OVERVIEW OF UK PREPAREDNESS

The future urban battle space will operate across a web of capabilities within the population, the first responders and the government. The strength of a nation's capabilities is a finite capability that must outlast and overwhelm terrorist networks that draw on in-country knowledge and in-country actors supported by global terrorist reinforcements.

An overview of the UK capabilities identifies a strong web of capabilities that are linked through a command and control system that operates symbiotically with major incidents and within the day-to-day relationships and communications between government departments. The structures are strong because they reflect continuous operational arrangements rather than the imposition of new systems during a time of crisis. The strength of response also comes from building on interpersonal knowledge of key actors between government departments and agencies and between government and business.

---

[2]Material regarding the United Kingdom is based on a paper by Sally Leivesley delivered at the NATO STS-CNAD Workshop, Portugal, 2006.

There has been a resistance to concepts of a terrorist czar and centralized operation of homeland security because the British culture tends to internalize and normalize the terrorist threat rather than acting as though it is a new phenomenon that requires a set of new structures. For this particular nation the strengths are cooperation, liaison, and a preexisting command and control that works through lead departments and Cabinet Office. The UK approach builds on the decades of counterterrorist experience with Northern Ireland and a history of successful peacekeeping work abroad where integration with local communities and the use of influence on "hearts and minds" of the public are key strategies. These strengths are not a model for other nations but can be judged as a successful adaptation of a nation to a long-term internal threat of national terrorism and a strong use of intelligence-led policing. Close relationships and cooperation in intelligence and police work has crossed boundaries and avoided some of the isolationist silo effect that is inherent in intelligence structures.

There are currently adaptations being negotiated for changes to the police force structure where plans have been foreshadowed to reduce the many county forces into a smaller number and to structure counterterrorist operations on a national level to make police intelligence work more effective. The use of resources is guided by threat assessments and a flexible allocation process so that the availability of police and other resources such as customs agents is matched to threat profiles of regions and activities. There have been significant plans for an increase in the numbers of persons recruited within the security services and a realistic understanding of the time taken to train and absorb growth rates of up to 50% over several years.

In preparation for the more serious risks from global terrorism, the UK emergency services had some advantages from work leading to the Commonwealth Games in 2000 which created lead capabilities in the first responders and assessments of chemical, biological, radiological, and nuclear (CBRN) protection technology for decontamination and detection. While some services benefited from this early work, the majority of the services commenced the focus on CBRN capability planning from 2002 onwards and the acquisition, testing, and exercising with equipment to protect responders and the population is ongoing. The national defense capability in CBRN protection has fed into civilian protection from a long and successful history of defense research into protection and detection. The extensive work in decontamination in Iraq after the first Gulf War and preparation of over tens of thousands of troops for CBRN protection prior to the Iraq invasion have also fed into the national capability.

British industry has supported the strength of the counterterrorism response by the emergency services with a focus on CBRN equipment and the innovative capacity of small and large teams within industry has contributed to covert and overt systems for detection and protection. The market for commercialization has not been significant in comparison to the USA and there is a move by US-based defense industries to open up companies within the UK. The need for joined-up research and development across countries is being publicly recognized by leaders in the defense industry (Sugar, 2006).

Significantly, the British government has also foreshadowed a rationalization of the security expenditures in the budget process to bring the cost of counterterrorist measures into a budget item that can be compared to other expenditures. There has been an acknowledgment that all ministries have a role in counterterrorism preparedness. There has also been recognition of the global links of terrorists and the close cooperation that is required internationally for security operations against terrorist financing and for investigations into individuals with multiple identity documents. The threat to the UK was described in February 2006 in terms of recent plots being of a different scale, with mass casualties, no warning, often suicide delivered, and with the potential for CBRN weapons. Al-Qaeda is described as a global terror threat with "hatred of our very existence." A commitment to enhancing services with equipment and training has been given by the government. Police forces had been increased by 16,000 persons since 9/11 including 6,000 more within the Metropolitan police. By 2008 annual expenditures for the Metropolitan Police are to be increased by £75 million, and £135 million is to be committed to regional intelligence. By 2008 the investment in counterterrorism and resilience will be £2 billion a year, which is double the expenditure before 9/11 (Brown, 2006).

4.1.2. TRAINING AND EXERCISES

Training and exercising have been part of the command and control systems of the emergency responders and over decades of operation with major incidents and national terrorism these capabilities have become formidable in the preparation of personnel for disciplined threat assessments, crisis response, and recovery. The systems are exported across the world with British trainers working in many countries in commercial sales of these systems. Diplomatically the use of British training institutions, particularly by defense, police, and fire services countries, has strengthened the ties across many national boundaries and created valuable personal links between persons at relevant rank positions for continuing cooperation.

The input of resources and solutions to the new forms of threat from global terrorism and in particular CBRN threats has been felt least by the local authorities. The focus of government planning has been to create a regional resilience structure and to create a joined up national response capability through regions which coordinate the local authorities. However, the Civil Contingencies Act (UK Government, 2004) has at the same time placed the burden of delivery of most services to the public onto the local level. These new responsibilities have not been enabled by budget allocation as the actions required by legislation to coordinate and deliver services at the local level were considered to be cost-neutral.

Local authorities have a difficulty that is common to all similar bodies in other countries where the significant risk of global terrorism creates a call for resources and technology applications that are expensive and technically difficult to assess. This causes difficulty with training and information to the citizens and to plan coordination of resources for catastrophic levels of attack. The higher levels of capability at regional and central government may coordinate resources in a spectacular attack but acquisitions policies, budget, testing and upgrading of the resources necessary to manage local casualties, and loss impacts are beyond the capability of local authorities.

Currently central government and the regional structures are highly active in establishing a strong coordination and in testing their capabilities to coordinate, and exercises are used to do this, including liaison with the USA in the TOPOFF (US Department of Homeland Security, 2005a) series, which was most recently held in April 2005. The UK exercise was called ATLANTIC BLUE and Canada also participated in the three-country test, which focused on incident management, intelligence and public information. The scenario was a simultaneous attack in Connecticut with a chemical weapon and in New Jersey with a biological weapon. This was the USA's first international exercise, requiring cross-border communications with Canada and the UK. While rated as generally successful by the Office of the Inspector General for the Department of Homeland Security, their report notes that the management of the incident with the new National Response Plan and the National Incident Management System created some stresses for the players. State and local participants also determined that the early involvement of federal and international assets was unrealistic (US Department of Homeland Security, 2005b).

The pattern of business response for preparedness has generally followed a focus on critical infrastructure protection after 9/11 when the criticalities of interdependencies of modern urban life became more obvious. Critical infrastructure, such as transport, telecommunication, and energy providers, are expected to coordinate as a second tier of responders behind the first responders at local levels within the UK. All relevant agencies have meetings with the local authority first responders, and join in training and risk assessment at the local level.

Small and medium-sized business involvement in contingency planning has followed this to some degree with a level of awareness, although not a set system of preparedness. The multinational business leaders have followed a global operation common to all international companies, but there has been some wariness about introduction of protective equipment and changes to work practices while the threat is undefined within the UK. Global companies have an option of risk transfer by moving production and trading to another country if one location is affected. This provides a solution that can affect any country, where a central financial district attack leads to a "flight of capital" to the home countries of any foreign companies in the attack zone.

The British public has been highly cooperative with police counter-terrorist operations and police requests for public assistance. Public cooperation translates into a flow of information to police on unusual activity and a confidence in and cooperation with police advice at any time of crisis. A close liaison between police and facilities managers ensures that proactive management of evacuation, and of shelter within buildings, is forthcoming. In London a pager alert system assists this process. Additionally some of the energy efficiency programs in London that create web sites for a company or general public links to share private vehicle transport provide an additional transport option for persons in an evacuation. This type of solution was used by the workforce during the bombing attacks of July 2005.

This general overview shows that the web of interactions and capabilities is a complex one, and that there is significant strength of input from the public response, as well as from the preparedness of the first responders, and the commitments of government to a strong leadership and to providing technical solutions as well as an exceptionally strong command and control.

4.1.3. CASE STUDY: LONDON BOMBINGS JULY 2005

London Bombings: July 7 and 21, 2005[3]

Three improvised explosive devices (IED) exploded on London Underground trains shortly after 8:50 a.m. on July 7, 2005 within about 50 seconds of each other. The fourth IED exploded on a double-decker bus in Tavistock Square, less than a kilometer from Kings Cross Main Line Station

---

[3]The raw data for the analysis in this section is drawn from two reports by Hazard Management Solutions, *London Mass Transportation System Bombings 7 July 2005, Triton Quick Look Report*, Updated July 19, 2005, original issue July 4, 2005; *London Mass Transportation System Attempted Bombings July 21, 2005, A Triton Quick Look Report*, July 22, 2005, www.hazmansol.com, UK. These two reports were produced as extended reports from an unpublished data base tool that records terrorist attacks each month across all countries. The gross database content covers over 33,000 incidents 2001–2006.

at 9:47 a.m., some 57 minutes after the three underground devices were initiated.

The trains were carrying approximately 800 persons and on two of the trains which were on the Circle Line, seven persons died in both the Circle Line explosions. On one Circle Line train the device exploded in a section of shallow sub-surface tunnel and blew through a dividing wall hitting a train passing in the opposite direction. The third train, which was on the Piccadilly line was in a more confined tunnel, about 70 feet (21.3 m) underground, with 25 persons dead and several hundred injured. This deep tunnel explosion damaged the structural integrity of the tunnel itself, which hampered rescue and forensic investigations. In all three trains the devices were initiated in the standing area within the double doors of the carriage and were located in the front carriage in the two Circle Line trains and in the forward section of the second carriage in the Piccadilly Line train.

The fourth device in the bus devastated the upper area of the bus killing 13 and injuring 30 or more. A large number of passengers escaped over the rear of the bus where the roof had peeled off from the force of the explosion.

The devices were estimated at less than 10 pounds or 4.5 kilograms of explosives. The explosives in the devices were described as "home made," and appeared to be consistent with home made explosives, such as Triacetone Triperoxide (TATP) or Hexamethylene Triperoxide Diamine (HMTD). These are extremely sensitive explosives that have been used by radical Islamic terrorists. On July 21, 2005 a bomb-making factory had been found by police with materials that appeared consistent with these explosives.

The four suicide bombers traveled down to London on the morning of the July 7 incident. They appeared to have traveled in rental cars down to Luton in Bedfordshire where they all boarded a Thameslink commuter train for Kings Cross Station. One vehicle was later found in the car park of Luton Central Station, and police took 14 hours dealing with the explosive devices and bomb making paraphernalia in its trunk.

The second attack came on July 21, with three IEDs delivered by persons intending to be suicide bombers partially exploding on trains traveling on the London Underground shortly after 12:25 p.m. A fourth IED partially exploded on a London double-decker bus approximately 60 minutes later, and this was left on the bus by an individual. The three attempted suicide bombings on the Underground failed to completely explode in each case, and the three bombers escaped in the confusion of the attack, although some were chased by members of the public.

In the attack on the Victoria Line train at 12:25 p.m. there was a small explosion in one of the middle carriages, with white smoke circulating through the carriage and into the adjacent one. Passengers noted a man wearing a rucksack that had split and they moved into adjacent carriages and evacuated at the station after pulling an emergency cord. The man ran off in the direction of a hospital and escaped.

In the second train at approximately the same time, on the Northern Line, in a carriage in the middle of the train, the perpetrator was standing in the middle of the carriage and a lot of smoke was seen coming from his rucksack.

In the third attack at 12:25 p.m., which was on the Hammersmith and City Line, a small explosion occurred and a man was seen lying with arms outstretched on top of a rucksack, face up and the rucksack was ripped with substances oozing out.

The fourth attack was at 1:30 p.m. on a double-decker bus where white smoke appeared, and damage to the windows was discovered. A rucksack was in the footwell near the rear right-hand seat of the upper level of the bus, which had been split open by the explosion with a substance coming out of it. The suspect who left the rucksack was seen on CCTV footage at 12:53 p.m., and he left at 1:05 p.m. Three of these devices were confirmed to be similar to the July 7 devices.

In one instance the confusion of train passengers about what was happening led to offers of assistance to the perpetrator, who was initially thought to be hurt. The eyewitness stated, "I turned around and there was a man lying with his arms outstretched in a Jesus Christ position, lying on top of a medium-sized rucksack, face up. He had his eyes shut and there was a puff of smoke coming from his bag.... His rucksack was ripped and had some gooey lard coming out of it... the man got up, I could see some copper wire showing out of the back of his T-shirt" (Woods *et al.*, 2005).

The configuration of these failed attacks and the July 7 bombings appeared to be a symbolic cross of fire, and this was related to a warning that, "The west will be crying north, south, east and west. London will burn in these four quarters," issued by al-Qaeda on Europe Organization website, June 2005.

Actions by First Responders

On July 7, there was some initial confusion in the control rooms of the underground rail systems because the three incidents were assessed as an operational power incident. Initial advice to the public through the media conveyed this message and for some period the responses were governed by this assumption. The chronology of response reported in the media stated that at 8:50 a.m. a report came from the Metropolitan Line control room in the City that said there was a loss of traction or a power surge. At 8:51 a.m. another rail line control room reported a possible explosion on the Circle Line train 100 yards into the tunnel from Aldgate Station. The London

Underground network control center (NCC), that controls the 12 lines and 270 stations, gave an order to move from safety code green to amber, instructing trains already in stations to stay put, and those in tunnels to proceed to the next station and stop. The entire tube network was suspended at 9:46 a.m. Buses were suspended in central London until late afternoon. Mainline railway stations closed for much of the day. Signs warned motorists traveling into London that the area was closed (Leppard, 2005).

According to newspaper reports, the police major incident contingency plan was activated just after 9 a.m. and alerts were sent to police stations, fire brigades and ambulance centers. The London Ambulance HQ turned its main operations center into a Gold Control and they started alerting hospitals. An example of procedures at one of the hospitals was described as a response that cascaded calls to staff via cell phones and pagers, roles were assigned according to the action plan, and laminated cards detailed specific tasks for nurses and other staff. Some physicians were positioned at the entrance to meet ambulances, and others managed triage and clinics (Leppard, 2005).

The nature of the incident was comprehended when emergency services personnel responded, and at the same time victims were emerging from self-rescue onto platforms. London Underground staff walked into the tunnels to the trains, and reported the carnage underground. About 350 of the wounded were ferried in ambulances and buses to four hospitals, and one hundred persons were detained in hospital overnight. There were 22 with serious injuries and a further 350 persons were treated at the sites. Fire services deployed 40 fire engines and 300 firefighters. Police officers from the Metropolitan and City of London Police responded, and police were flown back from the G8 Summit in Gleneagles (Leppard, 2005).

The operation escalated early to COBRA[4] level, with the Home Secretary convening a meeting and the Prime Minister (PM), who was hosting a G8 meeting at Gleneagles in Scotland, participating. The PM addressed the

---

[4]COBRA (as described in http://en.wikipedia.org/wiki/Cabinet_Office_Briefing_Room_A) is a coordination facility which is activated in cases of national or regional emergency or crisis, or during events abroad with major implications for the UK. The term COBRA is used both for the actual facility, and for the committees which meet there. Physically it is a secure suite of rooms containing a bank of telephone lines, fax machines, computer terminals, videoconference facilities, and other state-of-the-art communication equipment. Its purpose is to enable the prime minister, senior ministers and key government officials to obtain vital information about an incident and secure lines of communication to the police and other emergency services, army, hospitals, and all relevant branches of government. The chairmanship depends on the nature of the incident or crisis.

nation from Gleneagles in Scotland at 11:30 a.m., and addressed the nation a second time at 5:30 p.m. that day (Leppard, 2005).

Some parts of the London business areas evacuated staff, while others continued working with advice being given by police for persons to stay within buildings. The loss of underground rail transport caused considerable difficulties for persons moving out of London, and people used alternate options, including walking out of the city. In the Dockland's financial area there were long delays as people waited for water transport to be arranged.

Difficulties of access to the scenes of the bombings were compounded by three of the sites being underground. The bus site was accessible and provided early forensic information for the policing operations. The overall activity that happened on the day of July 7 tested the emergency response but the actual activity on the ground represented a web of capability based on plans such as traffic management plans for London and basic capabilities of the front line responders. Adjacent to the site of the bus bombing on July 7 was a building where a meeting of medical personnel was being held. They moved to the scene and assisted.

Due to the shortage of ambulances police moved some casualties on buses to hospitals.

There were over 700 casualties within the three trains underground and one bus site, so there were multiple arrangements to coordinate for rescue, alongside the recovery of 52 victims and the four suicide bombers.

The management of the forensics and the fatalities was expedited by a mobile mortuary that was erected over a number of hours in a close-by location. The location of this facility reduced time in postmortems, identification, and provided a caring service to the bereaved relatives. The recovery and forensics operation lasted for many days with teams working underground. There was a considerable commitment by the emergency services working in the difficult and confined tunnel rescue and in the longer period for the gathering of forensics evidence. Ventilation and safety were significant issues and there may be long-term health effects on personnel from working in confined space to do retrievals under these conditions.

Public Response

During the crises caused by the attacks on July 7 and July 21, the public attitude continued to follow a belief that if business as usual was maintained, then the terrorists' goal of creating fear would not be achieved. Seriously hurt and shocked victims were passive and the trauma of their recovery, the suffering of their families and of the bereaved have not been given a high profile since the incidents, although there were commemoration ceremonies.

The battle against public fear was won on the psychological front until the July 21 second wave attack occurred. Thereafter the understanding that terrorists would not be deterred has somewhat impacted the level of anxiety expressed by persons using the London transport infrastructure. A successful police strategy of overwhelming force was demonstrated on July 28 to defeat the possibility of a third terrorist attack on a Thursday—6,000 officers were deployed in a high visibility operation at rail and Underground stations and on buses to reassure commuters (Tendler and O'Neill, 2005).

Strategies in policing were highly successful in response to the London July bombings because a swamping of the underground and associated bus systems with maximum presence of policing and numbers of armed officers created a public perception of protection, which acted to remove any certainty from potential terrorist groups that a coordinated action would be successful at that time.

This public policing activity impacted on what was a psychological "shock wave" effect of the two sets of bombing attempts just 2 weeks apart. Although the second wave attack failed, due speculatively to a mistake in the preparation of the explosives materials, the potential impact was significant on the public perception of vulnerability. Hence the strategy of police action after July 21 to overwhelm the potential target area, and having a high profile policing operation on the third Thursday was quite a significant action.

The terrorist objectives appeared to be public casualties and economic damage from people becoming fearful of using transport. This was an economic attack on one of the world's leading financial cities, and did not work in deterring London workers. The public demonstrated a robust reaction and wanted to return to work and use the transport system, though there were fear reactions to persons on transport that were identified as culturally similar to the two sets of bombers on July 7 and July 21. This caused some offense, and in the longer term has been one of the negative effects on the Muslim population, but in general there was a unified community response. The Metropolitan Police worked very hard and in a high profile manner to ensure that all the faith communities were involved and consulted in activities after the bombing.

One of the real consequences of the July 7 attacks was the tunnel infrastructure, which was old and which had some failure points from the explosion overpressure and fragmentation effects. However, the disruption period to the lines was limited and the damage was recovered.

Post-July 21 Attack Actions

On July 25 in a press conference the Deputy Assistant Commissioner Peter Clarke, Metropolitan Police Service, Anti Terrorism Branch said that three of the men entered Stockwell underground stations just before 12:25 p.m. There were five devices recovered, with the fifth being found by a member of the public on July 23 abandoned in an open area. All of the devices had been placed inside dark colored rucksacks or sports bags and were made with the same type of food storage container (Metropolitan Police Service, 2005).

On July 27 a suspect was arrested with a Taser stun gun in Birmingham when police raided a house and found a suspect package, and cordoned off 100 houses in the area.

On July 27 in north London police recovered a large amount of chemical compounds from a lockup garage that could have been used to make explosives (BBC News, 2005). Rubbish chutes and bins of a nearby high-rise building had materials that could have been bomb-making materials.

On July 28 armed police launched a major operation in Notting Hill, West London and suspects were persuaded to surrender to police after some hours of siege on apartments.

The extensive police investigation is ongoing, and some of the statistics released on the level of activity show the resources required for a global investigation following an attack by small networked groups. In the July 7 investigations, 50 physical sites were searched, 11,000 statements were taken, 24,000 physical exhibits were logged, and 12,000 leads in total were to be followed up (Brown, 2006).

Conclusions

1. Catastrophic plans may need to be added to "credible incident" contingency preparedness because of the unexpected and devastating nature of attacks such as the July bombings especially if these are repeated as "shock waves" of attacks.

2. A reversal of the psychological drivers of being first responders may be required so that "response" is secondary to concepts of preemptive overwhelming of an attack. This requires some basic military concepts of battle planning rather than the civilian concepts of commitment of additional resources once there is an appreciation of the scope of the incident. Some first responder plans do incorporate this process.

3. Robust high capacity communications technology or a communications hub for first responder communications between agencies and for international operability is a challenge for all countries. Interoperability is

essential in building a nation's capability to pool resources and to share resources internationally.

4. Acquisitions policies of governments for first responders may need to be flexible and linked into defense acquisitions policies to maximize cross-applications for defense technology into the first responder market.

Energizing Public Capability

1. Recognition of public capability as a factor in overwhelming attacks may be an important frontline approach rather than a secondary measure. The psychology of waiting for advice and government response may need to be reversed to enable the public to survive the unexpected embedded homegrown terror attacks. This is a reverse engineering of traditional command, control, communications, computing, and information systems ($C4^I$). It sensitizes the public who are the front line victims to becoming responders who are able to recognize cues and exercise survival capabilities.

2. The reduction of casualty risk by energizing public response in the first 30 seconds and 5 minutes of an attack period can deny the terrorist objectives in causing fear and mass casualties. Self-rescue and assistance with serious trauma was the most important mode of survival for the victims of the July 7 bombings isolated in the underground rail network for some period of time before official rescue services arrived.

Cooperative Campaigns

1. Cooperative campaigns can be developed between business and government to prevent terror attacks and between first responders and the community during an attack. Global strengths may come from international police, security, defense, and political campaigns that are planned to overwhelm terrorism.

2. Campaigns can be embedded in normal operations rather than being a new set of systems and can run under the radar of terrorists. The more successful campaigns will be ones where terrorists cannot develop countermeasures, such as media psyops, or be able to monitor progress in systems for detection, or be aware of systems that are deployed to protect against weapons effects.

4.2. National Emergency Preparedness in the Netherlands: Uncertainty in Doing the Right Things[5]

4.2.1. CHANGE IN RISK AWARENESS

Homeland Security has become an important issue over the last 6 years. It started with the fear about the millennium causing our computerized society's technologies to collapse, which created a higher priority need for taking care of business and social continuity. After 9/11 the world realized that there was a change in orientation for disasters. "This could not happen to me" was changed to "This could happen to us."

After the bombing in Madrid, European political society started to look in the mirror and asked, "Are we prepared for major disasters or major terror attacks?" The Netherlands, a well-organized country, was "sitting on the sofa," because people thought that if they could beat the sea, they could beat everything. And then, one Saturday afternoon a fireworks factory in the middle of the city of Enschede exploded.

The uncertainty of doing the right thing is growing. A series of questions occurs to the people and the political leadership of the nation: how can we defend ourselves, are we prepared enough, what can we expect, and how can we manage everything? The Netherlands, a well-managed country, was burdened with a delicate management issue: How can we manage the unexpected?

4.2.2. THE DUTCH CRISIS CONTROL ORGANIZATION

After an extended period of inactivity in emergency management the Dutch government dismantled the crisis preparation forces and placed the responsibilities in the hands of the fire departments. At the operational level there is no longer a special disaster management operational team. To effectively manage a disaster the society turns to the same first responders who handle day-to-day emergencies: firefighters, policemen, health, and safety people and the military. In the Netherlands these forces are organized in 25 "Safety Regions."

The Netherlands distinguishes two levels of decision making during a disaster, terrorist attack, incident or other special occasion: (a) policy level and (b) operational level. The policy level mainly involves coordinating

---

[5]Material regarding the Netherlands is based on a paper by Jan Otten delivered at the NATO STS-CNAD Workshop, Portugal, 2006.

the 13 departments (ministries), 12 provinces, 25 regions and 467 munici-palities. When a crisis happens, the National Coordination Centre (NCC) coordinates the decision-making units and the Country Operational Coordi-nation Centre (LOCC) takes care of managing all the available operational people and resources. All the organizations have made arrangements and protocols with each other. To regulate information exchange the Ministry of Internal Affairs developed an Information Communication and Technology (ICT) system to ensure uniform communication. The operational level is managed by the 25 Safety Regions. During a crisis they work on a multi-disciplinary level. This cooperative approach is organized into 31 processes. These Safety Regions have all made a risk assessment priority diagram and equipped their staff.

In 2005 the Dutch government started a project for civil and military collaboration. Now the command and control management of the army is used to ensure a fast and confident way of operations (Figure 1).

Figure 1. The Netherlands operational chain

4.2.3. A SCIENTIFIC APPROACH TO EMERGENCY MANAGEMENT

Dealing with uncertainty is very difficult and creates diffuse reactions. How-ever, new studies and views can help the country managers to deal with the unexpected.

The first question: What can we expect and how can we organize this? Figure 2 indicates the analytical methodology used to approach the organi-zational question.

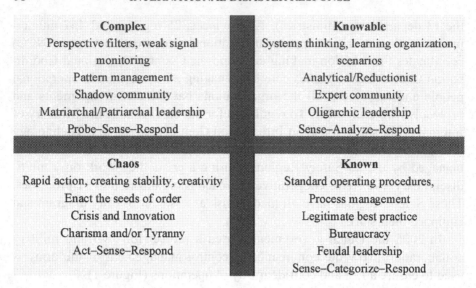

Complex	Knowable
Perspective filters, weak signal monitoring	Systems thinking, learning organization, scenarios
Pattern management	Analytical/Reductionist
Shadow community	Expert community
Matriarchal/Patriarchal leadership	Oligarchic leadership
Probe–Sense–Respond	Sense–Analyze–Respond
Chaos	**Known**
Rapid action, creating stability, creativity	Standard operating procedures, Process management
Enact the seeds of order	Legitimate best practice
Crisis and Innovation	Bureaucracy
Charisma and/or Tyranny	Feudal leadership
Act–Sense–Respond	Sense–Categorize–Respond

Figure 2. Snowden's Cynefin model: cultural sense making

The second question: The public asks for reliability and is that possible? Figure 3 indicates the relationship between reliability and flexibility.

The third question: How can we prepare ourselves? A block of Swiss cheese provides a visual representation of how societies undertake disaster preparation. Just as there are many routes for a mouse to run through the holes in the cheese, so there are many routes for people and organizations to take to become prepared for disasters. There is no single best route, as long as the goal of becoming prepared is achieved. Many paths and measures are

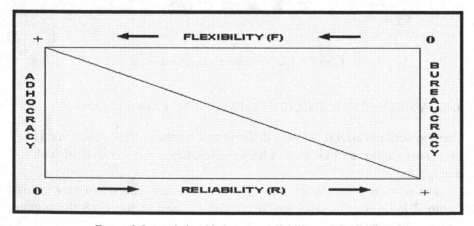

Figure 3. Interrelationship between reliability and flexibility

taken to avoid disasters. It is essential for a society to prepare its citizens, its first responders, and its social mechanisms to withstand natural, techno-logical, and human-caused disasters, including CBRNE.

The fourth question: Is it possible to use "network-centric" warfare for civilian network-centric operations? (For a complete review of "generational warfare" theory, see Appendix 2.) As noted in the OODA loop (Observation, Orientation, Decision, Action) conceptualization created by Colonel John Boyd, there is universal application for the decision-making process used by the military. The OODA loop theory is graphically displayed in a chart, which demonstrates the decision loops that characterize third-generation warfare. As a pilot, Colonel Boyd recognized that if he could transition through the four-phased loop consistently quicker than his opponent, over time he could overwhelm the opposition. First responders need to adopt this rapid decision-making style to succeed in asymmetrical warfare. (To view an OODA Loop diagram go to http://www.belisarius.com/modern_business_ strategy/boyd/essence/ooda_loop_sketch.htm.)

Ultimately the question is whether the nations can consistently transition through the necessary changes in adapting to terrorist attacks quicker than the terrorists can develop new strategies and tactics, yet maintain their culture. Nations are obliged to abide by the various treaties and conven-tions. Terrorists know no bounds.

4.3. Germany Civil Protection[6]

4.3.1. FEDERAL RISK ASSESSMENT AND CENTRAL EMERGENCY RESPONSE

In Germany, the local nature of disasters is recognized by assigning first response to local resources, usually from the Fire and Rescue departments. The federal states support the resources of the local authority for regional and supra-regional response with specialized units. The federal government provides protection against dangers which affect more than one area or entire nation, from natural disasters to complex technological or multisource emergencies.

Risk assessment in Germany has determined that the most likely sources of disaster are serious accidents, natural disasters, epidemics, damage of critical infrastructure, and man-made threats. With the fall of the Berlin Wall

---

[6]Material regarding Germany is based on a presentation by Heiko Warner delivered at the NATO STS-CNAD Workshop, Portugal, 2006.

the German government dropped war planning in the early 1990s. Recently, the possibility of an epidemic has heightened concern about the ability to respond to a nationwide onset of illness with existing medical resources. The recent development of a virulent form of bird flu has created a possible source for a future human epidemic, if the virus mutates to become person-to-person transmissible.

These civil protection scenarios need to be confronted by the central government or the federal states. NATO has been operating under Article 5 following the 9/11 attacks on the New York World Trade Center; so once again the German central government is developing a wartime orientation. New serious threats have been identified since the 9/11 attacks, such as serious nuclear, biological, or chemical (NBC) attacks, extensive fires, heavy storms, floods and fires, earthquakes, organized criminality, terrorist attacks, and military conflicts. German emergency management officials have noted that serious incidents are rising in the last 3 years, raising the number of missions assigned to the central government's emergency management arm. In the 1990s, there were 300 missions assigned to the federal emergency management agency versus 150 missions in 2005 alone (Werner, 2006).

4.3.2. THE LEVEL OF CAPABILITY OF FIRST RESPONDERS[7]

The typical training, equipment, and experience of first responders in Germany can be characterized by the situation in Hamburg. The State Fire Department and Emergency Medical Services of Hamburg have 5,000 employees, including engineers and HAZMAT technicians with special training for technological responses. However, they have not received special training for chemical, biological, radiological, or nuclear attacks. With biological and radiological attacks there may be time for planning an effective response, but chemical attack requires immediate protection of first responders as well as rescue and treatment of victims to be successful. Furthermore, in a site-specific attack with a chemical or an explosive device there is the additional concern that terrorists may have planted secondary explosive devices to target the first responders.

The Hamburg Fire Department has developed a protocol for using the victim reaction for evaluating the chemical family involved in the attack, For example, people with uncontrollable movements are most likely suffering from exposure to a nerve agent. These initial observations can also be used

---

[7]Material regarding Hamburg, Germany, is based on a presentation by Peer Rechenbach delivered at the NATO STS-CNAD Workshop, Portugal, 2006.

to determine the type and level of personal protective equipment that the first responders will need. In addition to their observations, the fire-fighters will use their chemical detection equipment to confirm their identification and begin medical treatment of the patients at the scene. They cannot wait for confirming laboratory tests if they are to save lives. For example, nerve agent requires immediate treatment with atropine and 2PAM to save the victim's life. Without the definitive laboratory analysis the firefighters have to make an educated guess, recognizing that a false positive will result in unnecessary and maybe life-threatening treatment, while a false negative may delay treatment until it is too late.

Dangerous substances include toxic industrial chemicals and weaponized chemical agents. However, dangerous concentrations vary. Those attacked in an open-air environment will benefit from the dispersal of the chemical into the ambient atmosphere, thus lowering the concentration below a lethal dose rather quickly. Weather conditions directly impact the speed of the dispersal, with sunny and windy days providing rapid dissipation, while cold, rainy days or nights engender slower dispersal of most agents. Since the "parts per million" measurement is crucial in determining the lethality of the dose, equipment must be adequately sensitive for each substance of concern.

While gas chromatography and other forms of laboratory-based wet chemistry provide highly accurate estimations of concentration, they take too long to save lives. The field instruments have limitations. Is the sensitivity of the field analysis system adequate for the type of substance being measured? Existing detectors for real-time detection have gaps in the type of chemical sensed. Furthermore, most require proximity to the substance to be tested, which may put first responders in peril.

Unlike a chemical attack with a rapid victim response to the attack, a biological attack offers no initial warning. Realization that an attack has occurred requires coordinated intelligence sharing among medical sources when illnesses first begin to appear. Possible sources of suspicion include increases in ambulance calls for service, public health reports of disease, and hospitals noticing increased admissions. Such information has to be collected to compare against a baseline for early detection of a biological attack. At present there is no good field detection system, although laboratories are using rapid analysis methods based on DNA, like polymerase chain reaction (PCR) and immunoassay systems. Coordinating the field intelligence and laboratory testing clearly requires the coordination of experts, systems and disaster response units when threats are detected.

The State Fire Department of Hamburg's solution was to create the Interagency Chemical Task Force supporting local disaster response units. They have special sampling procedures and equipment for detection of substances of concern. Their capabilities include a remote-sensing FTIR system, a mobile mass spectrometer system for the rapid identification of dangerous chemical substances, and a gas detector array for ongoing detection and sensing. Their team members include experts in a variety of applied sciences, such as meteorology and chemistry. There is now one team for radiological response, and four for chemical response. At present they are developing another task force for biological events that will include both first responders and technical specialists.

This beneficial system is not yet well known throughout Germany, as there is no National Incident Management System like the USA has, and no national disaster response system like Italy's. A system needs to be developed to inform local responders about the availability of this expert team, which is available to travel to other areas. The team is organized for rapid deployment, with a 60- to 90-minute response time. Transportation is available by army helicopter or federal police helicopter.

Hamburg's initiative is one way that a state is trying to fill a response gap in detection and analysis to enhance the fire service's life saving mission.

4.3.3. NECESSARY IMPROVEMENTS

During a terror attack with a weapon of mass destruction/disruption/killing the public has to be warned of the event, and advised on self-protection. During the Cold War most nations had civil defense sirens, but when the Cold War ended Germany removed the sirens due to the high cost of maintenance. In those days Germany was an "eastward facing nation," concerned only with a potential attack from the former Soviet Union and its allies. Now that terrorism is the threat, it is a 360-degree compass of surveillance and watchfulness. The current public alerting system rests on systems that are employed day-to-day for other purposes. For example, for small-scale emergencies the public address system in police cars is used to notify the residents of a community of danger. There is also a national emergency alert system based on radio and television announcements, but today few people are at home watching television during the workday. Therefore, these inexpensive systems are unlikely to provide adequate public warning during a catastrophic terrorist attack, although radio and television can be relied on for extensive coverage of the aftermath of an event. Again the London subway bombing provides a good case study of this.

German Emergency Management officials have evaluated their needs and are aware of the safety gaps that exist. The solutions, unfortunately, all come back to money. In order for the local, state, and federal agencies to pursue solutions to the recognized emergency response needs, they require a guaranteed and simplified source of funds for financing civil defense. Furthermore, all of the effort cannot rest with the government sector. Nations need to try public–private partnerships to develop community-wide response capabilities. For example, many firms already have emergency response equipment to deal with technological problems on their site, and agreements might be made to use this personnel power and equipment to augment public agency capabilities. Unfortunately, there are political problems in accepting private sector donations to government agencies.

For example, DaimlerChrysler wanted to donate cars to the German civil defense agency, but they could not accept them. The law, intended to prevent favoritism, requires that public agencies acquire their equipment through a formal bidding process. In the end the agency bought the very cars that DaimlerChrysler wanted to donate, using scarce funds to purchase equipment that could have been acquired for free (Werner, 2006).

More difficult is the need to develop improved specialty capabilities to combat biological weapons of mass destruction/disruption/killing. For example, there is a need to improve epidemiological capabilities to provide more lead-time for vaccinating the general population, or providing pharmaceutical prophylaxis. The problem is that modern epidemiology does not begin until a disease outbreak has been identified, which means they would be 5–10 days behind the attack, depending on the incubation period associated with the biological weapon, and the time it takes to recognize that an unusual number of cases have been identified. During traditional flu season the recognition of a biological attack might be further delayed.

In addition, civil rights and medical necessity come into conflict in the current procedures for isolation and quarantine. While the medical community may be able to diagnose the disease, the community's first responders have no procedures or legal protections to enable them to enforce population containment. In Germany you cannot use military units to keep the people confined (as an example, if authorities wanted to quarantine a stadium). Furthermore, trying to block exits would most likely lead to panic and crush deaths, as has been seen in soccer riots. How can we plan to keep contaminated victims or those already sick from spreading contagion among the rest of the population? Is this an appropriate role for the local police? How would they be trained? What kind of personal protective equipment would they be given? What civil rights would have to be suspended to enable the quarantine? Would local police officers be willing to use lethal force again their own community to enforce the quarantine?

Another gap is the lack of stockpiles of state-of-the-art materials needed for civil protection. Most European nations gave away all their shelter supplies at the end of the Cold War, so there are now no capabilities for population shelters in most nations. Government leaders know that it is not popular to spend money on such supplies, but emergency officials realize the need to develop new caches, but have to get financing from parliament.

Another capability gap is developing support staff and interagency units for terrorism response. Local agencies need to have staff available to support local or state or federal staff on the scene. Germany does not have interagency coordination. For example, they have 17 police forces of the states, six big government agencies, NGOs, and all the fire agencies that should be coordinated to respond to a disaster, but they have no system now.

What would happen if a 9/11-type event occurred in Berlin or Paris? Europeans are concerned that a 9/11-style event will be repeated. The government cannot protect all the assets, and a determined enemy seeking the weak spot will find it. For example, Europeans are concerned about safety surrounding the soccer world championship. It is reasonably easy to control the big stadium, but the concern is the uncontrolled crowds watching on big TV screens in large cities all over the continent.

The FIFA World Cup provided an example of cross-jurisdiction coordination in Germany. Traditionally there have been agreements between landers (subregional governments) to assist each other and share resources across boundaries. For example, the jurisdictions that border a river with a chemical plant adjacent to it would make joint plans to respond to a HAZMAT accident to protect the river and lessen downstream impacts. Make less high-risk targets more vulnerable. In the case of the World Cup, risk analysis was shared through a World Cup Coordinating Group. The same emergency managers from each lander met to discuss the World Cup and also have made the comprehensive plans against natural disasters. Among their strategies are the co-location of essential supplies at strategic locations when planned or anticipated events could lead to disasters. While the cache movement is probably not secret, and the terrorists know things have moved, even insiders do not know exactly where resources are, so just knowing about the activity probably does not help terrorists with their attack planning.

4.4. Protection Against Terrorism through Emergency Management Structures in Norway[8]

4.4.1. JUSTIFICATION AND RATIONALE FOR EMERGENCY PREPAREDNESS

This section presents national emergency preparedness structures in Norway and Denmark. The organizing and the elements of the responsive systems are identified, and their applicability for the handling of major terrorist

---

[8]Material regarding Norway and Denmark is based on a paper by Morton Bremer Maerli delivered at the NATO STS-CNAD Workshop, Portugal, 2006.

attacks is briefly assessed. In the second part of the paper, the findings and possible pitfalls and challenges associated with domestic and international emergency management is given. Here, a general appraisal of the differing dynamics and elements of intentional and non-intentional catastrophic events is, inter alia, given. Too often traditional accident mediation seems to be conflated with the management of major terrorist attacks. These scenarios may differ significantly in terms of timing, exposures, casualties, damages, media pressure, information vacuums, etc. Understanding these differences and taking them into consideration could be vital for proper emergency planning and responses.

4.4.2. NORWEGIAN EMERGENCY PREPAREDNESS: PRINCIPLES AND PRIORITIES

In a proposal for a Norwegian 2002 overall plan for national preparedness against means of mass destruction, the term "means of mass destruction" has been preferred to "weapons of mass destruction." The reason is that *means* is a wider concept, which comprises *weapons*, but which is more apt in a civilian context or in connection with terror and unintended incidents. Clearly, however, the expressions "weapons of mass destruction" and "NBC weapons" are in general use, and remain the basic terms in military contexts.

The expert group behind the plan defined the term "means of mass destruction" as follows: "Means of mass destruction is a term applied to radiological, biological and chemical substances that can potentially cause injury to persons or loss of human life on a large scale, heavy material damage, and the destruction of infrastructure. In addition, the threat to use as well as the actual use of such means is because of their nature likely to create great fear and concern among the population." However, the term "Weapons of Mass Effect" may be even more appropriate, taking deeper into consideration both physical damage and psychological impacts of such events.

The Norwegian emergency response is based on the following list of priorities for action:

- Life and health
- Environment
- Damage to property

The guiding structural principle is that the institution responsible for a public function under normal circumstances also carries the responsibility for emergency preparedness. Most government agencies at different levels, and to a varying extent several private organizations and enterprises, therefore have civil protections tasks and are responsible for preparing and implementing contingency measures in times of emergency. This also means that responses to unexpected events should follow the "LEON principle", i.e. be implemented at the lowest effective operational level.

4.4.3. NORWEGIAN LEGISLATION AND REGULATIONS ON EMERGENCY PREPAREDNESS

Legislation concerning Norwegian emergency preparedness may be divided into two pillars, consisting of Acts, prepared delegations of powers for law-making, regulations and directives. The first concerns the protection of the population in times of hostile acts and war, and the second relates to peacetime emergencies. The following Acts provide the main framework for administrative operations in times of crisis/war.

- The War Act of December 15, 1950, which contains directives regarding the extension of the powers of the Government and other administrative authorities.
- The Supplies Act of December 14, 1956, which is a power of attorney whereby the Government, in war or imminent threat of war, is given far-reaching possibilities for initiating supply measures.
- The Act on Civilian Defense of July 17, 1953, which contains directives regarding the organization, the civilian service duty and the civil defense duties of municipalities and private operators (e.g. shelter construction and secure drinking water supplies).
- The Act on Health Preparedness of June 23, 2000, implemented July 1, 2001, which establishes the duties of health and medical institutions in regard to taking all the necessary measures in peacetime to guarantee similar operations in times of war.

There is no singular Norwegian law dedicated solely to civil protection in peacetime. Many government agencies and private organizations do, however, have civil protections tasks and are responsible for civil emergency planning, and every part of the administration must ensure that necessary detailed emergency plans are put into effect. This is mainly regulated through laws and regulations concerning specific sectors. The following legislation is important in this regard:

- Police Act of August 4, 1995, which ascribes the police authority to implement emergency measures and coordinate rescue efforts
- Fire Protection Act of December 8, 2000, implemented as of January 1, 2001
- The Planning of Building's Act of June 14, 1985 amended November 24, 2000 with implementation as of January 1, 2001, which obliges the

municipalities to evaluate risk areas before regulating land to industrial or housing purposes

- Protection against Pollution Act of March 13, 1981, which contains directives on municipal preparedness and implementation of restraining measures against acute pollution
- Regulation of July 4, 1980 on the main principles for the organization of the Search and Rescue Services
- Regulation of September 21, 1979 with Instructions for the work of the County Governors in regard to civil emergency preparedness
- Regulation of December 12, 1997 with directives for regional coordination in peacetime emergencies
- Regulation of November 24, 1961 regarding Civil Defense efforts to avert or limit the consequences of natural disasters or other serious accidents in cases where the ordinary rescue services are unable to provide adequate resources
- Regulation of June 26, 1998 concerning nuclear emergency preparedness.

4.4.4. STRUCTURAL FRAMEWORK

The ministerial structure, and the responsibilities and organization of the ministries, determine the nature of the security and preparedness work done in Norway. That work is accordingly based on the following principle: "The authority responsible for a sector in peacetime is also responsible for the necessary measures to prevent damage, preparations for emergencies, and implementation of measures in times of crisis or war."

From this, moreover, it follows that the greatest possible similarity should be sought between how matters are organized for peacetime, crisis and wartime conditions, because the environments that perform social functions in peacetime are also best equipped to deal with extraordinary circumstances. Civilian emergency planning, therefore, needs to have a local basis, where locally defined risks, vulnerability, and needs direct the attention and actions of central government. National and regional authorities support the municipalities by:

- Improving the quality of the local emergency preparedness
- Improving the local competence for crisis management
- Increasing the awareness of the local political leadership to the possible consequences of natural and technological hazards

- Providing resources and coordination in an acute crisis situation that is beyond the handling capabilities of each municipality.

Municipalities, county municipalities and the state are obliged to draw up preparedness plans for the health and social services for which they are responsible. In addition, hospitals, water works, and the food control authorities are independently responsible for drawing up preparedness plans for their own activities.

Each ministry is responsible for the emergency planning and crisis management within their particular sectors. As a consequence, each ministry must organize its own crisis management team. Its organization will be based on the ministry's regular management structure. No permanent central planning group with regular members has been established for crisis management. Following the December 2004 tsunami catastrophe and massive critique of the governmental handling of it, this may change. Deliberations are underway for a joint crisis center.

The Ministry of Justice has special political responsibility for cocoordinating emergency planning and crisis management between the ministries and elsewhere. This includes responsibility for developing new national guidelines and ramifications, and to make principal decisions regarding the Norwegian civil preparedness system. The Directorate for Civil Defense and Emergency Planning (DCDEP) is the executive body of the Ministry of Justice, and is responsible for supervising the regional level with regard to civil emergency planning.

DSB shall renew, reinforce and unite the safety and emergency preparedness and response work. DSB shall also contribute to ensuring that the risk associated with private and public activities is continuously reduced through good health, environmental and safety work. The objective is to reduce the vulnerability of the Norwegian society. DSB's focus is on major accidents and other extraordinary situations, both accidental and deliberate.

DSB shall work to prevent loss of life and to protect health, the environment and essential public functions and material assets in connection with accidents, disasters and other undesired occurrences in times of peace, crisis and war. The Directorate shall have a full overview over developing vulnerable situations and looming perils which threaten society, regardless of whether they are related to accidents, disasters, or other undesired occurrences. DSB shall take initiatives to prevent such incidents from occurring, ensure that the necessary preventive measures have been taken, as well as that adequate preparedness is established so as to minimize the consequences of any undesired situations that may arise. In the event of inadequate safety and preparedness measures, DSB shall take the initiative to follow up with the responsible authorities.

DSB is the national public authority for municipal and inter-municipal fire services, the local electrical safety inspection authorities and the county governors' emergency preparedness and response work. DSB is also responsible for professional and administrative follow-up of the Norwegian Civil Defense, the Emergency Planning College, the Norwegian Fire Academy and the Civil Defense's three regional schools.

DSB reports to the Ministry of Justice and the Police. Its activities are organized around head office, 20 civil defense districts, five civil defense camps, five schools, and five regional inspectorates for inspection and control of electrical safety. The Directorate comprises a total of approximately 700 employees.

The Ministry of Justice also has the administrative responsibility for the Search and Rescue Service (SAR), and through the Police the operative responsibility for coordinating all search and rescue activities in any given emergency. Norwegian SAR has a two-tier command structure, with two main Rescue Coordination Centers (RCCs) at the top. Each has operative responsibility for the southern and northern part of the country, respectively.

The regional level consists of 19 counties, the County Governor being the highest representative of the central government. There are, however, only 18 County Governors, as two counties (Oslo and Akershus) are organized with a common County Governor. The County Governors are responsible for promoting emergency planning at the local level and participating in the planning of support of the military forces, as well as being responsible for environmental issues, agriculture, and the inspection of municipal administrations. The County Governor has by regulation the authority to coordinate and supervises all civil emergency planning within the county, and this authority increases greatly in crisis and war. In a major crisis the County Governor is responsible for operational coordination.

The 435 municipalities are required by law to undertake civil emergency preparations within certain sectors, such as Civil Defense, health, and resource allocation. There is, however, no general law obliging the local government to take an overall responsibility for civil emergency planning. The County Governor and the DCDEP, therefore, have to supervise the local level through information, motivation, and providing the municipalities with methods and tools for improving their civil emergency capabilities.

The municipalities have a broad range of political responsibilities, and the normal principles of responsibility will be maintained during a crisis. The local authorities may therefore be involved in handling aspects of the crisis such as policy decisions and information to the public and the media. Representatives from the local police, fire department, or medical service will act as on scene commanders according to the nature of the actions taking place. Normally the actions are coordinated from the local Rescue

Sub-centers (RSC) at the local police headquarters. There is one RSC for each of the nation's 27 police districts.

4.4.5. FIRST RESPONDER COMMUNITY IN NORWAY

Search and Rescue services in Norway rely on resources from public services, voluntary organizations, and private firms. The majority of the resources are contributed from the following government sectors:

- Police: 6,000 officers whose duty it is to coordinate operations
- Civil Defense: mobilizing force of 50,000, about 15,000 of these are trained for actions in peacetime disasters and some 2,000 are mobilized in small search and rescue groups acting on very short notice
- Public Health and Ambulance service: about 1,500 ambulance units available
- Municipal Fire Departments: 2,000 professional and 14,000 part-time firemen
- Armed Forces: about 5,000 personnel available
- Royal Norwegian Air Force: 10 Sea King long-range search and rescue helicopters, a number of Bell UH-1B helicopters, 7 Lockheed P3B Orion and other aircraft that can perform search operations
- Royal Norwegian Navy: frigates, corvettes or small crafts

As a result of an the allocation of NBC resources which took place in the autumn of 2001, the fire services of nine of Norway's biggest cities and towns[9] now have equipment available for the detection of chemical warfare agents. They also have detection equipment for industrial chemicals.

The Norwegian Ministry of Defense foresees a comprehensive quality reform for the Home Guard (Forsvarsdepartementet, 2004). The reform entails role specialization and improves the quality of training, equipment and reaction capability. The Home Guard's ability to carry out its missions in a more complex and unpredictable threat environment—including counter-terrorism in support of civilian police—will be significantly enhanced.

There are, furthermore, a number of other public services with important resources and competencies in the event of disasters. Voluntary organizations also play an important role in search and rescue operations. The Norwegian Red Cross has between 300 and 400 rescue teams, with a total of 17,000 individual members. The Norwegian Society for Sea Rescue has about 40

---

[9]Oslo, Bergen, Trondheim, Stavanger, Kristiansand, Drammen, Bodø, Tromsø, and Romerike (Lillestrøm/Gardermoen).

vessels stationed along the coast. Besides these, there are a number of other voluntary organizations and several private firms that contribute with important search and rescue resources. A more detailed overview of Norwegian WME Emergency Preparedness Equipment is given in Exhibit 1.

EXHIBIT 1: NORWEGIAN WME EMERGENCY PREPAREDNESS EQUIPMENT AS OF MAY 2002 (CONTACT, 2002)

Preventive Measures

Shelters: A total of 2.7 million places, 50% of which are in shelters providing collective protection against means of mass destruction (NBC protection).

War removal and evacuation plans: Plans of this nature and practical exercises based on them have given Civil Defense units insight into, and experience of, moving large groups of people.

Protective masks: Stored under Civil Defense auspices are 300,000 gas masks for the civilian population, 100,000 field masks for Civil Defense personnel and 17,000 protective bags for children. This protective equipment is intended for use when moving out of contaminated areas. It is thus not designed for longer stays in such areas. Plans exist for the replacement of all protective masks over an 8-year period.

Detection

Nuclear preparedness detection equipment: The Civil Defense has control of 156 "Automess" instruments for measuring radioactivity. A further 16 "Automess" instruments are held by the organization's decontamination units. Capable of measuring very low levels of radiation, the instruments can measure alpha-, beta-, and gamma-radiation. In cooperation with the nuclear accident preparedness organization, the Civil Defense carries out measurements of background radiation once a quarter at a number of places in Norway. The measurements are carried out by regular personnel and radiation measurement patrols. The radiation measurement service is currently under review.

Detection equipment for means of chemical warfare: Each of the 16 mobile decontamination units in the Civil Defense is provided with a hand-held meter for chemical warfare agents. The Civil Defense also possesses detection powder and powder for the detection of gas in liquid form, as well as M256 equipment for showing the presence of gas.

Alarms

The Civil Defense currently has about 1,250 operational alarms for alerting the civilian population. An event involving means of mass destruction may be made known through the signal, "Important message—listen to the radio." Given a better budget, more alarms can be made operational and the alarm system can be improved.

Damage limitation

Regular officers and personnel: To combat nuclear and chemical means of mass destruction, the Civil Defense has a force numbering 11,000–12,000 officers and other ranks capable of carrying out first aid and safety assignments. The peacetime contingency teams are the best trained and can be mobilized at the shortest notice. They number about 2,500. To reinforce and relieve them, the Civil Defense has a total of 6,700 officers and other ranks in other contingency teams.

Decontamination units: The Civil Defense has 16 mobile decontamination units intended for on-site decontamination of persons who have been exposed to radiological and chemical contamination. Each unit has detection equipment for means of N and C warfare—Automess and CAM (chemical agent monitors), 16 suits for protection against chemical warfare agents, and a tent to stay in after decontamination. This equipment is identical with that procured by the Board of Health for use at hospitals.

Cleansing agents: The Civil Defense has plentiful supplies of the adsorbent Fullers' Earth for absorbing gas in liquid form.

Other front-line resources: Civil Defense units are equipped with modern firefighting, rescue, communications, and medical materiel. A fleet of vehicles is also available. Nationwide alarm systems and shelters are maintained with a view to a preparedness situation.

Firefighting and rescue material

Firefighting and rescue materiel includes about 900 fire pumps and roughly 600 kilometers of fire hose, motor saws, air bags, jacks, generators, and extraction equipment.

Medical equipment

Chiefly portable first-aid equipment comprising of infusion solutions and supplies, oxygen, splints, bandages, stretchers and blankets.

Communications

Lightweight radio and cable communications that can be linked in networks. The equipment makes communication possible with the police, the fire services and other rescue organizations.

Special resources for preparedness against means of mass destruction

- Protective masks for Civil Defense forces: about 100,000
- Protective masks for the public: about 300,000

- Protective suits for Civil Defense forces: about 2,000
- Protection bags for children: about 17,000
- Detector paper for means of chemical warfare: about 4,000 pads
- Adsorbent, Fullers' Earth: about 90,000 packages
- PDRM 82 intensitometer: 80
- Radiac meters, Automess 6150 AD 1: 156 (plus 16 earmarked for the mobile decontamination units)
- Radiac meters, contamination, NM20: 95
- Hand-held meters for chemical warfare agents (CAM): 16 (earmarked for the mobile decontamination units)

EXHIBIT 2: NORWEGIAN NUCLEAR EMERGENCY PREPAREDNESS

The National Preparedness Organization for Nuclear Accidents, established by Royal Decree of June 26, 1998, is built around the Crisis Committee for Nuclear Accidents. The Crisis Committee relies on a set of scientific advisers, recruited from major government agencies and institutions with special expertise in nuclear accident preparedness. The County Governors represent the regional component of the nuclear accident preparedness.

The Crisis Committee has the responsibility for development of the continuous nuclear accident preparedness. In the acute phase of a nuclear accident, the Committee shall determine the action to be taken; in the subsequent phase, it will act as advisor to the authorities. The main object of its work is to protect life, health, the environment and important social interests. The Crisis Committee is also responsible for coordinated information to relevant authorities as well as media and the public.

The Norwegian Radiation Protection Authority's tasks in a crisis are to assist the Crisis Committee with scientific expertise, information gathering, situation assessment, collation of measurement results, and so forth. In more normal circumstances the Radiation Protection Authority is the contact-point for international warning agreements and maintains round-the-clock alertness. It is affiliated to national and international networks for measurement of radioactivity in the air.

The Norwegian Radiation Protection Authority has its own preparedness unit in Sør-Varanger, close to the Russian border. There is an extensive Nordic and international collaboration in the preparedness area with joint exercises and various development projects.

On February 15, 2006, a new Royal Decree was issued, substituting the 1998 Decree. By this, the responsive structure was renamed "The National Preparedness Organization for Nuclear Emergencies"—"reflecting the current threat picture, as well as legitimizing the actual tasks of the emergency preparedness structure" (NRPA, 2006). The new name reflects a broadened mandate the nuclear emergency preparedness now includes events involving a range of radiological and nuclear incidents, in addition to the traditional accident response. However, no major organizational changes were made.

4.5. Danish Emergency Preparedness

4.5.1. PRINCIPLES AND PRIORITIES

The principal objective of the Danish emergency preparedness is to prevent, reduce and remedy injuries to persons and damage to property and the environment in case of accidents and disasters, including war or the imminent danger thereof. The emergency preparedness rests, like the Norwegian one, on the principle that the agency normally carrying a certain responsibility, continue doing so during times of crisis.

The mission of the national rescue preparedness has been laid down on the basis of the political intentions for the preparedness, that is, the Preparedness Act and the political preparedness agreements concluded at any time. The mission is worded as follows: "The national rescue preparedness is to safeguard the population and society against accidents and disasters in the best possible way. The rescue preparedness authorities shall thus work to minimize the harmful effects of accidents and disasters and increase their preventive efforts in an attempt to avoid the occurrence of accidents and disasters."

Following the 9/11 terrorist attack on the World Trade Center, the Danish government decided to reinforce emergency preparedness in Denmark. Among the initiatives was the creation of a research pool, with the aim of collecting and comparing data from larger accidents and catastrophes and conducting research into the operational response necessary in the event of a terrorist attack. Other initiatives included new anti-terrorism laws, expanded recourses to the two Danish intelligence services, and new emergency preparedness equipment (Dalgaard-Nielsen and Hamilton, 2006, p. 11).

4.5.2. DANISH LEGISLATION AND REGULATIONS ON EMERGENCY PREPAREDNESS

The Preparedness Act is the statutory basis of the activities of Emergency Management Agency (DEMA) and the national rescue preparedness. According to the act, the rescue preparedness shall comprise the national rescue preparedness including the national regional rescue preparedness and the municipal rescue preparedness. The 2004 Danish Defense Force Act stipulates, moreover, that the armed forces should assist civilian authorities in the event of terrorist attacks, disasters, or accidents.

Denmark, like Norway, participates in the international cooperation on combating terrorism, among other things through the UN Security Council Resolution 1373 and the EU framework decision 2002/475/RIA on the combat of terrorism. Within the framework of this cooperation, Denmark has an obligation to prevent the preparation of terrorist actions and to ensure that Denmark is not a "safe haven," in terms of terrorists having the possibility of setting up bases from which terrorist actions can be planned and implemented. Furthermore, Denmark is committed to ensure the prosecution of and suitable sanctions against terrorists, as well as to assist other countries in investigating criminal cases regarding terrorism. The Security Council resolution and the framework decision together with the UN terrorism financing convention, among others, provide the basis of the "anti-terror package" passed by the Danish Parliament, Folketinget, on May 31, 2002 and becoming effective as from June 8, 2002.

4.5.3. STRUCTURAL FRAMEWORK

The Danish system has three levels, where local actors—again—are meant to react first in a case of emergency. Local forces are, if need be, backed by, regional preparedness centers. The handling of more extensive incidents are managed and coordinated by regional staff. If local and regional forces are overwhelmed by one or more events, emergency structures on the national level will be triggered. A new National Operative Staff chaired by the National Police Commissioner will be charged with coordinating the responsive efforts (Dalgaard-Nielsen and Hamilton, 2006, p. 16). This national entity is composed of representatives of the national police, the Danish Defense Command, and DEMA.

Two key civilian partners at operate in parallel at the national level. Whereas the National Police reports to the Minister of Justice, DEMA—originally reporting to the Ministry of Interior—now reports to the Ministry of Defense. As part of the Preparedness Act, which came into force on January 1, 1993, the former fire service and civil defense were integrated into one single-strand rescue preparedness service to be used in peacetime as well as during a crisis and in war.

The Danish Emergency Management Agency (DEMA) is responsible for the national rescue preparedness in Denmark. DEMA also handles supervision and counseling tasks related to municipal rescue services and other authorities, general development in the field of preparedness and a series of operative tasks. DEMA is an agency under the Ministry of the

Interior and Health, and handles relevant rescue preparedness tasks related to the ministry. Moreover, DEMA is in charge of coordinating the planning of the civil sector preparedness and carries out tasks concerning civil sector preparedness not allocated to other ministries. Finally DEMA is the Danish national safety authority and responsible for the Danish nuclear emergency preparedness.

The Security Intelligence Service (PET) regularly evaluates the terror threat against Denmark, and other national authorities follow its recommendations. Last autumn, PET established its National Contact Group for Counterterrorism, an expanded partnership between PET and public authorities. This year, PET has started a similar contact group for the business sector.

4.5.4. FIRST RESPONDERS IN DENMARK

The national rescue preparedness has a staff of some 600 persons. About 170 of these are employed in the central Emergency Management Agency. The rest are employed at the Agency's seven rescue centers and three schools. The national rescue preparedness in Denmark comprises the regional divisions of the national Rescue Preparedness Corps. These divisions, stationed at six of the seven rescue centers (one of the rescue centers does not perform conventional rescue preparedness tasks), assist the municipal rescue service and other preparedness services upon request. There are 275 municipalities in Denmark. The rescue centers, inter alia, train conscripts for rescue preparedness and to handle the State's operative regional tasks as part of the overall rescue preparedness—that is, mainly by assisting the municipal rescue services. Furthermore, all seven rescue centers are in charge of regional emergency planning.

Each year, the rescue centers train 900 conscripts who serve for 3 months. During this time, among other tasks, they complete a qualifying firefighting course. Furthermore, the rescue centers train 500 conscripts who serve for 6 months. These conscripts complete a rescue course and a course on HAZMAT as well as the firefighting course. The trained conscripts thus constitute a good recruiting basis for the municipal rescue services or a private rescue corps.

The regional rescue centers are staffed round the clock. At the request of, for example, the municipal rescue services or the police, the rescue centers provide assistance at fires, as well as for rescue and environmental operations. The rescue centers can respond with special equipment and personnel in force at short notice. The rescue center personnel also constitute the basis of DEMA's international operations. Concerning vehicles, machinery, and other equipment, the regional rescue centers function as centers of expertise, each with its own specialty. The purpose of these centers is thus to gather and further develop professional knowledge, mainly in logistical fields, to the benefit of the overall rescue preparedness.

In case of accidents and disasters occurring with prior warning, including crises or war, calling up a national mobilization force can extend the preparedness. This force can be built up gradually as needed—basically because the force is divided into two parts. The first part consisting of 500 men must be operational within 48 hours; the second part consisting of 4,500 men can be called up gradually and will require a longer period of preparation.

4.5.5. NORWEGIAN AND DANISH EMERGENCY PREPAREDNESS COMPARED

In both Norway and Denmark, the emergency preparedness has been continuously developed and adapted to the changing demands made by the society and the changes in the security-policy situation. Ideally, this should ensure a preparedness level that is capable of intervening swiftly and flexibly in response to all types of accidents and disasters. In Denmark, emergency measures are defined and formalized by a dedicated Preparedness Act. In Norway, the activities rest upon a set of laws, regulations and Royal Decrees.

Structurally, Norway, and Denmark have chosen the same three-level approach, with emphasis on local (first) response, and keeping ordinary responsibilities and chains of commands even during crisis. This would, normally, allow for swift and qualified responses—backed by redundancy and enforcement at level two or three, depending on the scale and the magnitude of the event. Regular exercises both at the respective levels and in between the levels, ensure streamlining and correctives for the complete organizations in both countries. High-level political interest in emergency preparedness measures seems to be present in Denmark, as well as in Norway, resulting in pressure on the respective agencies to orchestrate and coordinate their efforts (Dalgaard-Nielsen and Hamilton, 2006, p 12).

In both countries, however, there seem to be tendencies toward domestic "mission creep" and emergency response turf battles between agencies. Indeed, the Norwegian Ministry of Defense views terrorism "as a form of warfare." Accordingly, "acts of international terrorism will have a national security dimension that threatens state security in addition to societal security" (Strat, 2005), hence giving the armed forces a more prominent role operationally even in the domestic civilian sphere.

Whereas Norwegian emergency planning and preparedness do rely on some military support, civilian-military ties seem to have developed further in Denmark, both formally and operationally. The Danish Ministry of

Defense now administers, through DEMA, cross-governmental civilian preparedness and response planning (Dalgaard-Nielsen and Hamilton, 2006). However, the Danish police, and hence the Ministry of Justice, retain responsibility for the domestic operational coordination in case of an incident—accidental or intentional—involving responses from more than one governmental body (e.g. police, firefighters, and health personnel).

In Norway, civilian protective and responsive activities are to a larger extent managed under the same ministerial roof, the Ministry of Justice, through its two "operational branches": the Police and the Directorate for Civil Protection and Emergency Planning. The benefits are obvious. It allows for one joint line of authority from central to local level within the areas of fire, rescue, and general preparedness. Operationally, it establishes a common professional sphere covering prevention and preparedness for incidents at central, regional and local levels.

Such an organizational structure, however, rests critically on updated knowledge about relevant emergency preparedness allocations, procurement, equipment, and competence governed by other ministries. While the inter-governmental coordination and information activities may benefit from such an arrangement, optimal uses and integration of, for instance, civilian and military recourses may suffer.

The US concept of "Homeland Security" may, furthermore, have become more deeply rooted in Danish domestic protective security endeavors, at least judging from academic contributions (Dalgaard-Nielsen and Hamilton, 2006). Still, "Homeland Security" has not been ingrained in the Norwegian emergency preparedness vocabulary.

EXHIBIT 3: TERRORISM RISK ASSESSMENTS

Terrorism countermeasures have traditionally included political governance, socioeconomic measures, communications and educational efforts, military inter-ventions, judicial and legal measures, and law-enforcement and intelligence activity.[10] In the USA, measures against terrorism have been put together under the rubric of "homeland security," whose strategic objectives (in order of priority) are to: prevent terrorist attacks within the USA; reduce US vulnerability to terrorism; and minimize the damage and recover from

---

[10]To identify and classify preventive and counter-terrorist measures, the Terrorism Prevention Branch of the United Nations Office on Drugs and Crime has developed a set of eight categories from a "Toolbox of Measures to Prevent and Suppress Terrorism." A detailed summary description of these may be found in Alex Schmid, "Statement at the Inter-Agency Co-ordination Committee Meeting on the Illicit Cross-Border Movement of Nuclear and Other Radioactive Material," International Atomic Energy Agency, Vienna, May 26–27, 2003.

attacks that do occur (Office of Homeland Security, 2002). These measures are all likely to play a role, small or large, in fighting future acts of terrorism. Their use, efficiency and costs will differ significantly, however. The respective countermeasures must be scrutinized and prioritized accordingly.

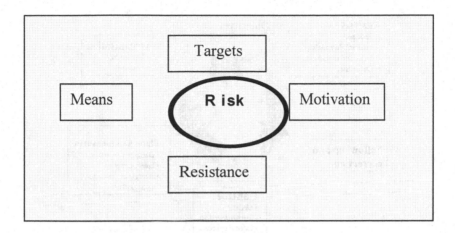

Figure 4. Simple terrorism risk model

Here, a simple terrorism risk assessment model, considering the different elements of the terrorism threat, could prove useful. At the outset, it should be acknowledged that terrorist abilities rest on a combination of intentions and capabilities. Highly motivated groups possessing the proper means to perform terrorism represents the highest risk. Equivalently, low motivations and low capability would represent a low nuclear terrorist probability, and hence a low risk. In the complete absence of either motivation or means, the probability of nuclear terrorism would become zero.

Similar considerations could be made for societies' vulnerabilities to terrorism. "Vulnerability" may be regarded as the (societal) ability to withstand injury, or its ability to recover from damage. Operationally, vulnerability could be defined as "a function of the reciprocal multiplicative relationship between risk and preparedness" (Hedge, 1987). Society is most vulnerable when risk is high and preparedness is low; by the same token, it is least vulnerable when risk is low and preparedness high. An absence of targets, or a high level of resistance, could reduce the risk of terrorism.

Hence, as reflected in the US Homeland Security strategy, terrorism threats, and thus proper countermeasures, have to be continuously monitored against terrorists' motivations and means, and against the resistance of the society and possible targets (see Figure 5).[11]

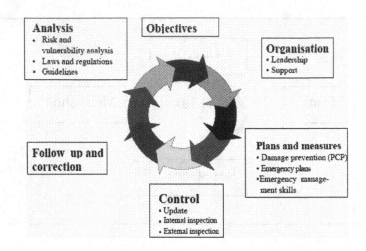

Figure 5. A model for a systematic approach to better preparedness. (From Norwegian DBS 1998/1999.)

4.6. Emergency Preparedness in Portugal[12]

Civil Protection in Portugal is organized on three levels: national, 18 regions, and 308 local services. In Portugal all disasters are local, and the mayor is the first official responsible for civil protection. Communities have Emergency Operations Centers (EOCs) and emergency plans with scenarios to drive the opening of the EOCs based on political and technical decisions, and the advice of the safety services. Portugal has links with the EU, NATO, and other regional groups.

Portugal has a variety of first responder agencies that participate in emergency response. These include the fire service, Security Forces which include

---

[11]The risk assessment model is taken from the presentation by Dr John-Erik Stig Hansen at the Conference "Homeland Security: Bridging the Transatlantic Gap," sponsored by The Danish Institute of International Studies, the Johns Hopkins SAIS Center for Transatlantic Relations, and the Norwegian Institute of International Affairs in Copenhagen, September 19–21, 2003.

[12]Material regarding Portugal is based on a presentation by Alvaro Campeao delivered at the NATO STS-CNAD Workshop, Portugal, 2006.

the national police, national republican guard, local police, armed forces and maritime service, the national medical emergency services, the Portuguese Red Cross, and other public and private social and health care services. All first responders were trained on WMD indicators before EURO 2004.

The Portuguese Civil Emergency Planning Council, which is responsible for crisis and warfare events, has to debate the definition of emergency versus crisis to set priorities. They have adopted a National Crisis Management System that assigns roles from the prime minister through the government. This system guides the decisions on the level of execution and level of support for any given event. It outlines decision levels and the chain of command, and describes relations with the media.

The Civil Protection agency maintains a national map of vital and sensitive points, which must be kept highly confidential. It is an essential element of planning for security.

Since the terrorist threat has been a concern, Portugal has hosted a variety of high profile events. These include Expo 98 in Lisbon and World Gymnestrada in Lisbon 2003 with big closing ceremonies drawing a crowd of 60,000 people. The UEFA Cup in 2005 drew 60,000 people, and also "Rock in Rio" in Lisbon drew 100,000 people. The Civil Protection agency continues to be concerned about planning major events in this era of terrorism.

For EURO 2004, England, Spain, France, Germany, and Italy worked together. Altogether 10 new stadiums all over country have been built, with a capacity of 10,000–60,000 in each. Terrorism prevention started with the design and building of the stadiums, including the development of internal and external emergency plans. There were special security concerns for two stadiums within 2 kilometers of the Lisbon airport. As a result there was good security inside the stadiums, but outside of the stadium it was harder to have security, and an attack would have gotten major media attention.

Preparation for the sporting events provided a platform for developing good cross-border cooperation among EU partners. Intelligence centers were linked with EOCs, and information developed during the sporting event planning was shared with other countries that had events. There were concerns about hooligans, but they were stopped in their own countries.

Medical preparedness is harder to develop and maintain. Hospitals are not well prepared for terrorist events. There is a little commerce in biological materials that requires special security, resulting in rather limited emergency preparedness in case of a biological terror attack. However, since there is no nuclear power plant within Portugal, there is little medical preparedness for dealing with the consequences of a radiological or nuclear event. First responders have detection devices in the field for biological and chemical hazards, which would allow some lead time to prepare hospitals and medical reserves to care for victims.

Safety is important for large events like the sports games, but planning must also include security. Three months before EURO 2004 there were violent attacks in the Atocha Railway Station, Madrid. It was originally thought that there was only one bomb attack, but then multiple devices were found. Spain was prepared because of its long-standing battles with the Basque separatist movement, ETA. Portugal worked with the Spanish explosive experts to learn as much as possible about the management of the attack.

However, even with this real-world example it is still hard to overcome the apathy factor, which suggests that "this won't happen in Portugal" (Campeao, 2006). Therefore on March 11, 2006, Civil Protection presented new planning scenarios combining all the possible threats. Internet sources of information were used, and the material details how to prepare and protect first responders. It is presumed that this will help in raising the concern about terrorism and transmit it to the citizens, bringing them into partnership with the government.

4.7. The US Response to Catastrophic Urban Disasters[13]

Catastrophic Urban Disaster events pose a challenge for first responders based upon the type, scope and complexity of the event. Events that require extensive search and rescue may include earthquakes, floods, hurricanes, tornadoes, avalanches, construction accidents, explosions, and acts of terrorism involving structural collapse. Initial actions by first responders are critical. Proper understanding of the value and importance of methodical search techniques require a disciplined and focused response, which is not always found during disasters (Schapelhouman, 2006).

In the chaotic, confusing and difficult world of initial response, proper search methodology can often take a back seat in the unfocused rescue effort. Overwhelmed and often emotionally driven responders may have poor training, minimal search equipment, and lack the overall discipline and experience needed to focus their initial efforts. A wide variety of external influences may further impact the situation (Figure 6).

Proper search methodology starts with determining the geographic parameters of an incident, followed by establishing an alphanumeric gridding

---

[13]Material regarding United States Urban Search and Rescue is based on a paper by Harold Schapelhouman delivered at the NATO STS-CNAD Workshop, Portugal, 2006.

system needed to organize the scene into smaller workable sections. Finally, effective deployment of specific search resources to prioritized targets, based upon tactical objectives can be achieved. Effective search is often the tipping point between time and lives saved. In the critical first hours and days of any catastrophic emergency, time is the key factor. Methodical search practices can bring organization to chaos, focus to confusion, and control to an undisciplined response.

Figure 6. Search and rescue

4.7.1. PREPARATORY PHASE

The successful response to a catastrophic event is often dependent upon what was done prior to the event. Mitigation may include planning, training, and purchasing of necessary items in preparation for a variety of emergencies. For managers and planners, the real challenge is often how to justify pre-event preparations, such as investing in mitigation strategies and measures, paying for training and equipment, and maintaining a progressive synergy over years or decades, when an event does not occur.

Once a jurisdiction or agency has determined that a specific risk exists, some form of mitigation or preparation should occur to minimize the effects. Once an event occurs, different strategies and skill sets are required to directly address the event. This may be complicated by a loss of infrastructure, personnel, and capability. It may be further deteriorated by a lack of understanding of who is in charge, what the chain of command will be, and the proper procedures needed in time of crisis.

Local responders can also become emotionally impacted or overwhelmed, because the event is occurring in their town—potentially affecting their families, friends, and even co-workers. Couple this with multiday events, fatigue, depression, system difficulties, or failures, and the stage is set for what often differentiates a disaster from a catastrophe. A system failure or collapse may occur—causing a "disaster within the disaster."

4.7.2. COMMAND AND CONTROL

In recent years, the USA has adopted the National Incident Management System (NIMS) as a national guideline of standard practices and principals, which clearly addresses system-wide issues involving command and control needed for dealing with emergencies under the National Response Plan (NRP). This system is based upon successful practices used in California under the Standardized Emergency Management System (SEMS) which uses the Incident Command System (ICS) as its cornerstone of system hierarchy, and an expandable template used for both first responders and elected officials to seamlessly interact and manage a crisis. (Department of Homeland Security, 2006; California Governor's Office of Emergency Services, 2006)

Key to this system is training which outlines system use, standardized position descriptions, roles, responsibilities and checklists, as well as a straight forward guideline process for physical resources to become typed, or categorized, as part of a standardized resource ordering and management tool (Figure 7).

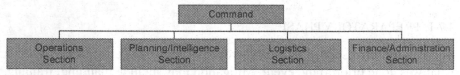

Figure 7. Incident command chart

4.7.3. PHASES OF RESPONSE

The US National Response System under the FEMA which manages the national Urban Search and Rescue (US&R) program identifies four primary phases of response to catastrophic incidents for first responders (California Fire and Rescue Training Authority, 2000).

- **Phase 1:** Most important in the first few minutes of any event is an understanding of exactly what has occurred so that the most acceptable course of action can be utilized. This is often called "size-up." Critical to a complete and thorough size-up, is establishing situational awareness and conducting some form of reconnaissance (recon) to access the overall extent of damage so that the most prudent actions, based upon the extent of the event and available resources, can be determined. This is often easier said than done.

Ideally, a singular field command post should be established and utilized to manage the operational elements of the event until a more permanent Emergency Operations Center can be opened or established. In some instances, based upon the size of the event and number of agencies involved, an Area Command Post consisting of representatives from all involved parties can be established to provide a cohesive and coordinated management team for the incident (Figure 8).

Figure 8. Command post

As critical as situational awareness can be, it is equally as important to quickly request additional assistance. It is imperative to understand the methods in which those requests must be channeled to be accepted and addressed, as well as specifically what is needed to help prioritize immediate needs. Responders should focus on developing a scene safety and control plan which consists of perimeter access control, immediate medical treatment and/or triage, movement of victims to a safe refuge, feeding and sheltering and a hasty search plan. Rapid removal, or sheltering in place as many victims as possible, is based upon the scope of the event.

A concept of operations (con-op) should be established to clearly outline incident priorities, safety issues, security issues, work to rest cycles, resource tracking, situational status, reassignment, integration of requested resources, and a clear understanding of who will be doing what, where, when and why. The con-op should be reviewed at least every 12–24 hours to monitor changes and address new and developing issues. This planning cycle and subsequent distribution of information to all managers through the chain of command should be developed as quickly possible.

Incident managers should establish an alphanumeric gridding, or area search system, to break the event into smaller more controllable and reportable sections. This will also maximize and prioritize existing resources. This process helps to focus resource ordering along with assignments. If controlled and utilized correctly, it will reduce freelancing and eliminate unnecessary duplication of searches.

- **Phase 2:** Most important to this phase of the operation is a methodical and systematic search for any living victims who may be trapped or unable to evacuate. This controlled process is called a primary search and requires a collective approach and focus to be successful. It relies heavily on a clear feedback loop between commanders and field responders. Proper methodology used during this phase would be to incorporate information gained from existing situational awareness and reconnaissance. A clear set of operational priorities must be complemented by broad strategies and implemented by specific tactics shared with all personnel. Significant issues such as hazard assessments, structural stability, and overall operational safety all impact the selection of priorities and corrective actions.

  Use and prioritization of the 9-1-1 caller information and/or specific tracking of cellular phones should be quickly evaluated, and acted upon. This type of information is invaluable, but offers a limited window of opportunity.

- **Phase 3:** Victims identified in this phase require special assistance. The capability to extricate, evacuate, or rescue them from the event is required. This may be a time consuming, high risk, labor intensive and technical operation. Such operations must be carefully scrutinized because they may result in catastrophic failure. A basic "reward realized" versus "risks to be taken" model should be developed and used to guide critical decisions. Those decisions will have to be balanced against available resources, political atmosphere, emotional restraints and time.

- **Phase 4:** This phase represents the human recovery phase which may require a significant secondary search of all affected areas and structures. Specific to this phase should be the elimination of high-risk nighttime operations, rotation of responders, and revision of the risk versus reward matrix. Clear protocols for how the respectful search and recovery of deceased individuals should be managed must be developed. The need for improved debris removal, increased safety, and the integration of heavy equipment should all be evaluated.

Groups of local first responders should be rotated out of the event so that they can deal with any impacting personal issues involving family, friends or personal property. All responders should be exposed to some form of critical incident stress debriefing, peer support, and follow up medical screening.

Time is the enemy of the Search and Rescue Team, and may also factor into victim recovery. Survivability of victims varies from even to event, but can follow a distinct pattern if studied over a number of events.

EARTHQUAKE COLLAPSED STRUCTURE SURVIVABILITY MODEL: (CALIFORNIA FIRE AND RESCUE TRAINING AUTHORITY, 2000)

- First Hour   91%
- First Day   81%
- Second Day   36.7%
- Third Day   33.7%
- Fourth Day   19%
- Fifth Day   7.4%

The effects of serious medical conditions, known as crush and compartment syndrome, may further complicate the ability of the rescued victim's survival based upon the timing, type, and quality of medical intervention.

This model of survival expectations may need to be adjusted for other types of disasters. During both the Oklahoma City bombing and the World Trade Center terrorist attacks, there were no survivors located after the first 12 hours. In contrast, one of the last survivors of Hurricane Katrina was an elderly man who was found trapped in his home 19 days into the event as the waters receded in New Orleans (California Urban Search and Rescue Task Force 3, 2005).

4.7.4. SEARCH CONCEPT OF OPERATIONS

While the development of a Search and Rescue Concept of Operations is critical to the success of the incident, most managers know very little about successful search operations. While most models prove the need for well-trained citizens, first responders and technical rescue teams, an overall appreciation or benefit of proper search methodology may be lacking in the initial hours and days after an event.

EARTHQUAKE RESCUE MODEL: (CALIFORNIA FIRE AND RESCUE TRAINING AUTHORITY, 2000)

- 50% of all victims either self-rescue or are assisted by convergent rescue volunteers
- 30% of all victims are rescued by a combination of convergent rescue volunteers or local first responders from non-structural entrapment
- 15% of all victims are rescued by local first responders and specially trained technical rescue teams from structural entrapment
- 5% of all victims are rescued by specially trained technical rescue teams from significant structural entombment

The concept of operations for a terrorist or flooding event may look very different than that of an earthquake. Based upon other impacting circumstances, the ability of victims to self-rescue, or to help others, will be limited.

While most organized search and rescue teams would have you believe that they understand the value of methodical search, many spend more of their time, money and emphasis in training of technical rescue skills. Incidents such as the Oklahoma City bombing and World Trade Center attacks at times deteriorated as convergent volunteers, local first responders and even well-trained technical rescue teams utilized five-gallon buckets to remove debris by hand, layering down debris in hopes of finding victims. While significant finesse is a requirement for these types of operations, the line between "dig therapy" and true benefit can become blurred at times, as victim survivability slips away.

Additional survivability time can be lost if emphasis is placed upon the recovery of bodies, or human remains. These types of operations can also become controversial as normal methods of recovery must be modified through the use of disarticulation, or parting, due to deceased individuals being pinned by debris. During the Oklahoma City bombing and World Trade Center incidents, technical rescue personnel utilized disarticulation methods to free deceased victims only to be criticized by shocked local officials who could not comprehend the depth of technical expertise needed to locate, access, and recover these individuals outside of local norms.

In contrast, during Hurricane Katrina local officials in New Orleans were criticized by the media because dead bodies were floating in the water, and left on elevated overpasses. Such reports gave the perception that the event was out of control. Few understood that in many cases rescuers had tied floating victims to fixed objects and used global positioning devices to mark their coordinates so that they could later be recovered. This allowed rescuers to focus on the immediate task of saving live victims and not using valuable time conducting body recovery.

Recent developments in the use of computers and mapping software used in conjunction with global information satellite imagery along with geo-spatial data can help responders obtain real-time vertical imagery of events. In order to maximize the use of this type of information, it is recommended that events be divided into alphanumeric grids allowing for a systematic and prioritized search process. If done correctly, this process can reduce or even eliminate redundant searches, provide an additional coordination and reporting tool, and help to create a more disciplined search methodology and focused rescue operation.

4.7.5. THE ROLE OF SPECIALISTS

Search Team Manager

The Search Team Manager must be familiar with urban search problems and priorities for a variety of impacting events. Key to this individual's role is the development and implementation of the actual search plan. The Search Team Manager functions as the search/reconnaissance team supervisor, sketches and records information, and or communicates details and recommendations back to the Incident Commander or Task Force Leader to implement real-time decision making.

Technical Search Specialist

The Technical Search Specialist utilizes electronic visual and acoustic/seismic listening devices along with other methods to successfully work under the developed search concept of operations and within the designated search plan and grid.

Canine Search Team

The Canine Search Team consists of one trained and certified dog and a skilled and experienced handler. The team can be used for either search, or body recovery, but not always both, due to the number of specific skills the animal and handler must posses. Canine teams must be certified for urban disaster environments, and redundant verification and alerts using multiple teams is required prior to implementing specific rescue-related activities. Weather conditions, fatigue, and site hazards can limit or prevent canine use and probability of success percentages should be determined daily to gauge the true effectiveness of canine teams.

Technical Search Equipment

The advancement of new search equipment which is well beyond the basic acoustic/seismic and electronic visual formats continues to enhance the area of urban search. New developments in the areas of thermal, infrared, ultraviolet and electromagnetic wave detection radar sensing will greatly improve search procedures and probabilities for success.

Developments in the area of robotics, specifically unmanned aerial vehicles coupled with video and photographic capabilities continue to improve situational awareness and forward intelligence gathering and prioritization. Robots can work in rubble, debris, underwater and in normally uninhabitable environments. Robots continue to improve yearly as costs are reduced and benefits become more evident for incidents such as a dirty bomb or a flood.

4.7.6. SPECIAL SEARCH OPERATIONS

High-Risk Search

High-risk searches may involve structural collapse due to an earthquake or a terrorist event in which the potential for a secondary collapse or a secondary device may impact the first responder immediate decisions, actions and the safety of personnel. High-risk search environments may require unique mitigation techniques which need to be quickly improvised in order to be successful and timely. Adapting to changing environments and conditions should be regularly discussed. Training by responders should utilize scenario-based exercises to monitor acceptable protocols and desirable patterns of response. Risk versus reward scenarios must be discussed and decisions should be made based upon the best interests and safety of responding personnel, and victims.

Large Area Search

Large area urban searches may involve multiple structures, horizontal and vertical population density, impacted infrastructure and overhead and underground factors such as railways, highways, subways, and waterways. Priorities and objectives must be balanced against existing resources and capabilities. Additional resources will need to be quickly requested, mobilized and brought into the area to shore up a potentially overwhelmed local response.

Significant command and control challenges may be experienced in this type of environment due to security, support, logistic, communications, planning, and coordination challenges. Development of redundant capacity or capability may be needed, should local responders be otherwise engaged or equipment destroyed.

Flood Search

Urban flood search presents a variety of safety and logistical challenges. Prioritization should be based upon a systematic aerial site evaluation to determine size or area, water depths, navigation hazards, potential improvised landing zones and immediate search and rescue challenges along with needed resources.

Population support and movement may impact search and rescue operations if victim recovery outpaces the ability of resources to support and move moderate to large populations. These types of operations may detract from critically needed search activities in the most severe of conditions which will require a building to building methodical search process which can take up to three times as long as normal ground based search. Secondary searches must be done as water levels recede and opportunities to provide a more detailed, ground-based search are presented.

4.7.7. CONCLUSION

Comprehensive and successful search methodology is a practiced technique which requires planning, forethought, training, proper equipment, and operational discipline for the urban environment. A combination of tools and skill sets is needed to rapidly adapt, improvise and overcome a variety of situations and to compensate for various potential events.

The greatest challenge will be to respond during the initial phases of the event, when the potential for the human-created "disaster within the disaster" may occur. To overcome that secondary condition, a system's response must be adapted, utilized, and practiced. An understanding that a methodical, systematic search for victims must be accomplished prior to the onset of prioritized rescue activities will be crucial to the overall success of finding salvageable victims.

4.8. National Emergency Prepardness in Case of Terrorism in Russia[14]

4.8.1. EMERCOM ORGANIZATION

"EMERCOM is subdivided into several departments. These include the Department for Protection of the Population and Territories; the Department for Disaster Prevention; the Department of Forces; the Department for International Cooperation; the Department for the Elimination of Consequences of Radiological and other Disasters; the Department for Science and Technology; and the Management Department. EMERCOM also contains several specialized commissions and boards to coordinate and implement certain tasks. These include: the Interagency Commission of the Russian Federation for Fighting Forest Fires; the Interagency Commission of the Russian Federation for Floods; the Interagency Maritime Coordinating Commission for Emergencies on the Seas and Water Basins; and the Interagency Commission of the Russian Federation for the Certification of Rescuers. Working through the office of the Prime Minister, the Ministry can ask for private or Ministry of Defense or Internal Forces assistance. That is, the Ministry has international coordination power and the ability to tap local resources if required" (Thomas, 1995).

The main principles, structures and circuits of the EMERCOM organization in Russia respond to extreme situations, including terrorism. These systems enable the use of scientific and practical research to increase the efficiency of counteraction to modern terrorism. The structure of the "National Plan of Liquidation of Consequences of Acts of Terrorism" was created as a result of the application of the research which demonstrated that a separate approach was required for success against modern terrorist acts.

4.8.2. OPERATIONAL PLANS WITHIN EMERCOM

In Russia the territorial agencies maintain emergency response units and facilities that are used against terror attacks. Response begins with the emergency commission of the local administration, the civil defense and emergency management control unit. They are deployed when an emergency message is received by the local administration or executive agency, which then orders its civil defense and emergency management services to deploy

---

[14]Material regarding Russia is based on a paper and presentation by Boris Potapov delivered at the NATO STS-CNAD Workshop, Portugal, 2006.

the local response units. They start taking emergency and rescue actions. If their local resources are overwhelmed, they turn for assistance to their respective regional centers of EMERCOM of Russia that may dispatch additional manpower and facilities to the emergency area. Their resources include units of the Civil Defense Force, search and rescue squads, medical personnel, special machinery, means of communication, and similar resources for managing and mitigating the attack.

In case of a large-scale disaster, or a disaster of a unique nature, the information about it is received by EMERCOM of Russia and analyzed by the Crisis Management Center. This Center consists of coordinated resources of federal, regional and field/local-civil defense and emergency management personnel. After the analysis the response is carried out based on the plan-based algorithm of actions. The units and facilities of the central government join the local resources in the operation, including the units and facilities of other ministries and departments, with state financial and material resources involved.

In case of a disaster abroad, including a terrorist attack, if a foreign government or the United Nations Department of Humanitarian Affairs asked Russia for help, the Government of the Russian Federation makes a decision as to the form of assistance. This may include sending rescuers, medical personnel, other staff, equipment and means of transportation, delivery of food, medicines and other supplies for survival, financial support, and other logistical support. EMERCOM of Russia, as the emergency management ministry, is commissioned with fulfilling the corresponding tasks. So EMERCOM, in cooperation with the Ministry of Foreign Affairs of the Russian Federation and the National Corps of Emergency Humanitarian Response start an emergency response operation abroad.

For rendering assistance to foreign states, EMERCOM of Russia has such units as the Central Airmobile Rescue Unit, the Expeditionary Hospital, the LEADER Special Risk Rescue Operations Center, the Motor Column and the State Unitary Air Enterprise, all being parts of the Russian National Corps of Emergency Humanitarian Response. Should the need arise, state reserves are employed for sending humanitarian supplies to the destination point on aircraft and trucks of EMERCOM of Russia.

The effectiveness of emergency response depends on many factors. The most important of them are information regarding situation status, situational analysis, preparedness and rescue services, financial and logistical support, and the development of operational plans.

Situation status is crucial to mounting the best response to any disaster. Informational support comes from state bodies and the civil defense and emergency management staff, which includes their obtaining and exchanging of the needed data. Next, a system of situation analysis is applied, resulting in data to support making decisions at different levels of government regarding the selection of the best forms and methods of response, the mobilization of the necessary manpower and resources, and the activation of plans for interdepartmental and territorial cooperation.

Implementation of the decision to response to the disaster is supported by the existence of organized personnel trained for the emergency and rescue services. Programs are in place to create and maintain qualified responders, to ensure the availability of modern machinery and special equipment, to organize the means of transportation and a system of communication, and to provide in-field support units for the emergency response units and the affected population. Appropriate financial and logistical support of emergency and rescue operations is maintained according to the national plan. The analysis leads to a deployment plan to ensure the optimum nature of specific emergency and rescue actions and organizational activities in the disaster zone.

4.8.3. MANAGEMENT ACTIONS

Each of the above-mentioned activities and phases requires specific management actions, facilities and resources that may be united into several main technical blocks, as detailed below:

Informational Block

This includes activities aimed at obtaining information, using primary data about the scale and features of the disaster, and integrating the data needed in the course of the emergency and rescue operation. It determines the order of obtaining, exchanging and processing the information. Monitoring, communication and data-processing systems are used to support the response phase.

Commanding Block

The command element is responsible to use the developed information analysis to make effective decisions regarding the use of all resources available for life saving and rescue operations. It includes emergency planning and undertaking of emergency relief activities and their management, financial and logistical support of the operation, and coordination of the activities of various supporting agencies. It also includes the activities at the final stage of an emergency and rescue operation, such as analysis of its outcomes and issuing of recommendations. During the activities in question, computer systems and experts' estimations are used.

Mobilization and Resource Block

This operational arm of the response determines the selection, within a short time span, of the optimum composition of the response units and logistical support, in accordance with the nature of the rescue operation, and the order of employment of material and technical resources (their nomenclature, number, and implementation).

Transportation Block

It analyzes the organizational and technical ways of ensuring of rescue units' mobility and promptness of their arrival at a disaster site. They also ensure the delivery of the needed amount of equipment for rescue work and personnel support in the field. It is based on an inventory of the means of transportation: planes and helicopters, special purpose vehicles, river-craft and maritime vessels, aero-sleighs and cargo-chute systems.

Search and Rescue Block

This operational activity is comprised of the activities of rescue units and their support units aimed at limiting loss of life and rescuing the victims. Their tactics include localization of disaster foci, search for casualties and removing the danger to their lives and health, rendering first medical aid to living casualties and their delivery to hospitals, ensuring safety of the rescue operations personnel and other persons at disaster sites, taking actions aimed at recovery of an affected area and preventing the situation there from deteriorating. The logistical support for these activities includes many complex systems and interrelated units.

Medical and Sanitary Block

This public health-oriented unit is organized to provide the victims and other persons in affected areas with effective medical aid, preventing infectious diseases caused by deterioration of the sanitary situation, and averting the danger of such diseases spreading beyond the border of affected zones.

Life-support Block

This logistics-oriented unit uses a set of measures aimed at creating appropriate power supply, physiological, ergonomic (human engineering) and other conditions for autonomous life-support and effective work of the rescuers, and survival of the affected population. The facilities used are mobile camps, power plants, water extraction and treatment devices, food, sanitation, and hygienic supplies.

Operational Control Block

This unit determines the order of the actions of rescue units and their personnel in a disaster area and at specific sites within the affected zone, and their cooperation with support organizations and the units of the commanding block.

Rehabilitation Block

This unit begins its work as the immediate response to the disaster is resolving the life safety issues. Its work includes the whole range of activities aimed at withdrawal of the commanding units, rescue services, machinery and equipment to their points of departure and initial condition, the physical and psychological recovery of participants of an emergency response operation, and the resupply of their resources.

Each of the above-named blocks, apart from their sequence of actions, their composition, staff and the technical means involved, determines the requirements for the qualification of its personnel, the equipment it uses, the duration and quality of the work in question and the safety rules. All this must provide for achievement of the maximum result with the minimal time and material expenditure.

4.8.4. ACTION PLAN

One of the factors appreciably influencing the effectiveness of a rescue operation is its starting time. The research pursued by the EMERCOM of Russia shows that if the initial search and rescue operation can begin immediately and be achieved within the first hour after the disaster, the loss of life can be reduced significantly. For example, comparing rescue work that begins at the sixth hour after the disaster with rescue work that begins within the first hour after the disaster, a 30–40% decrease in loss of life can be achieved. In addition, if a twofold increase in the tempo of the work can be achieved, this results in a 35% increase in the number of the victims rescued alive.

All the mentioned technological blocks influence the duration of an emergency, and the start time and speed of the rescue work. Therefore, all of them determine the effectiveness of emergency response and, finally, the success or failure of a rescue operation.

As outlined above, the important condition of the increase of efficiency in the reaction time to terrorist threats requires the cooperation of not just all first responder agencies, but also of the public. Public education regarding terrorism is aimed at creating a "State of duty service" on the basis of telephone number "01." With a view to increase the efficiency of actions at liquidation of consequences of acts of terrorism, the following *Action Plan* is implemented:

- Coordinate the common approaches to the development of plans action for the first responder forces involved in liquidation of consequences of acts of terrorism, including the development of uniform methods to estimate the degree of threats and possible consequences from acts of terrorism.

- Establish uniform degrees of readiness for the forces involved for counter-action to acts of terrorism, and the order of actions necessary to maintain various degrees of readiness.

- Precisely define the powers and sphere of the responsibility of the head of works on liquidation of consequences of acts of terrorism.

An adequate and efficient response to acts of terrorism requires the development of special plans of liquidation of consequences of acts of terrorism. These plans must address the following questions:

- What are the possible adverse scripts of the succession of events, caused by acts of terrorism?

- What kinds of resources are needed and what is the best structure for the required staffing and resources to carry out the most effective search and rescue operations, field level medical care, and other urgent work?

- What is the best command and control system for the integrated responding forces, the order of their interaction, and the division of their spheres of responsibility?

- Considering that the terrorists may be indigenous to the area they have attacked, what is the order of preliminary creation and rational accommodation of necessary resources for operative reaction to various variants of terrorism response situations, including the means of protecting the population and first responder forces working in the centers immediately after the terrorists have been defeated?

- How do you manage the control of the unaffected population that comes to the danger zone to seek relatives or check on their property? How do you provide them with effective risk communication? How do you keep the rescuers apprised of the circumstances and condition of the community?

4.8.5. THE BESLAN SCHOOL TERROR ATTACK

The actions of the rescuers from EMERCOM of Russia at the Beslan school attack and murders were an example of the problems associated with the practical implementation of EMERCOM's counterterrorist operations strategy under severe real-world conditions. On September 1, 2004 terrorists attacked a school on opening day, capturing not only children and staff, but also family members who were present for first day of school festivities. The local government response was immediate, but the terrorists had stockpiled munitions and explosives previously within the school gymnasium where they kept the captives as hostages.

The Chechen rebels were protesting the presidential election in their homeland in July, an election that they claimed was rigged.[15] The men were armed with assault rifles, and the women wore suicide belts. EMERCOM of Russia organized the response to the school based on their existing plans. After 2 days of a standoff one of the Chechen munitions apparently exploded on its own, resulting in an assault by the Russian commandos who had been surrounding the school.

At 3:40 p.m. on September 3, 2004 the standoff ended as the gym roof collapsed and victims fled in a rain of bullets exchanged by both sides. The Russian commandos entered through a wall of the gym to rescue the children and their families held inside, and apprehend or kill the terrorists, regardless of gender. By the end of the siege there were 330 people dead, including 156 children and 26 terrorists. 700 people both inside and outside the school were wounded. The EMERCOM responders cared for the wounded at the scene, and took them to the hospital as quickly as resources permitted. By 9 p.m. they were clearing mines and munitions from within and around the school.

This tragedy confirmed the effectiveness of many of EMERCOM's plans, but also offers lessons for improvement, such as the strengthening of the command and control function in order to improve the coordination of the various response elements. The Duma is working on a draft law for a revised national security strategy.

4.9. Response to Unconventional Terrorist Attacks in Israel[16]

The high frequency of terror acts has forced Israel to change to a rapid method for managing the injured and cleaning up the crime scene. Cleanup is begun within 10–15 minutes after the attack. One benefit is that most of the victims are rescued so quickly that they survive. The responders are now being asked to consider how they will deal with even more catastrophic

---

[15]For details on the attack see J.F.O. McAllister and Paul Quinn-Judge, "Defenseless targets," *Time*, September 5, 2004.

[16]Material regarding Israel is based on a presentation by District Commander Yossi Segiv delivered at the NATO STS-CNAD Workshop, Portugal, 2006.

events on a larger scale. They have become accustomed to dealing with car and bus bombs where the casualties are limited to about fifty, and their protocols are based on this patient load. This Israeli model may provide some schema for use by NATO member countries when they experience a truly catastrophic event. "It is only a matter of time before Europe is attacked" (Sagiv, 2006).

Terrorism is the definition of a chaotic event, but first responders have to try to put order in it. Israel has no experience with CBRN-caused events, but years of experience with escalating explosives-based events. Therefore its terrorism response plans have to be evolutionary, based on conversations with first responders and academics to create the most effective methods. Field based plans then require soldier/officer input from the practical units that manage the field events.

4.9.1. HOMEFRONT COMMAND

Israel is protected against terrorism internally by a division of the Israeli Defense Force (IDF) called the Homefront Command. The IDF is divided into four commands: north, central, south and Homefront; and five districts: north, central, southern, Tel Aviv, and Jerusalem. The police have the same district zones. The Homefront Command was created in 1991 after the Gulf War, and its focus is civilian protection. During ordinary times the police manage ordinary crime and criminal activity. The Homefront Command provides a bridge between ordinary police responsibilities and army activities. It is built with a strong capability for research on population protection rather than war fighting, and although it is a part of the army, the Homefront Command staff work with the civilians.

The Homefront Command soldiers attend schools and receive education in dealing with emergencies. Each student has 1 week in school with a female soldier teaching emergency response. Among other responsibilities they over see the construction of houses in Jerusalem, where every house is required to have a sealed room for protection from possible chemical munitions attacks. Homefront Command soldiers oversee the planning approval for these specialized shelters within the homes. Homefront Command soldiers also provide training for local authorities on how to act in emergencies.

During times of war, their numbers are augmented by reserve soldiers who are retrained every year. The Homefront Command works closely with the civilian government, the police forces, and the local medical community in developing terrorism response plans.

4.9.2. TERRORISM RESPONSE PLANNING AND TRAINING

Terrorism response planning uses a variety of scenarios to develop both strategic and tactical response plans. Scenarios include chemical terror attacks, radiological attacks, anthrax envelopes, and biological terror attacks; while in Israel nuclear emergency planning is conducted by a different organization. Detailed plans and training are driven by intelligence analysis of the likelihood of an event, and its likely consequences. Analysts consider the enemies that are likely to act against Israel, the intelligence on their capabilities, and the likelihood of their acting in any given time frame. The plans cover the impacts of such attacks, the probability of attacks using these instrumentalities, and an estimate of the response cost for each type of incident. This analysis helps to determine what needs to be included in medical stockpiles.

For example, the probability of a chemical attack is high because it is simple to do. Chemicals can be produced readily, or bought as industrial chemicals, as shown by the Aum Shinrikyo attack against the Tokyo underground using home-brewed sarin. Conversely, producing and delivering a biological agent is difficult, and radiological material would have to be stolen.

Terrorist use of chemicals including warfare agents must be anticipated. Such attacks are lethal, and come without early warning. This means that it is highly likely that first responders will be injured because they do not wear personal protective equipment (PPE) all the time, although they have it in their vehicles. However, because it is hot and cumbersome they do not use it for routine calls, which may expose the first responders to the chemicals until the cause of the problem is recognized. Chemical attacks will cause great damage in crowded places. It is not likely that terrorists would use chemicals in open areas, because it is too difficult to get adequate concentrations to be effective, but indoors they would be used to great effect. Enclosed places yield more damage.

Chemicals cause immediate damage and mass casualties. The use of poisons and lethal agents against unprotected civilians would result in large numbers of casualties. Planning for such an eventuality has been going on since the first Gulf War. Israeli civilians have gas masks in home, supplied already in 1991. Many of these masks are no longer effective because the seals have deteriorated, and the government has tried to collect them back, but people will not give them up. There is a concern that the masks create a false sense of security in the population. The problem is that even a small amount of a chemical agent can cause great damage. The response plan calls for immediate medical treatment of the casualties, evacuation of the contaminated areas, and thorough decontamination.

The organizations involved in the chemical attack response include police, fire, Red Star of David, local authorities, health organizations, environmental department and Israeli Defense Force. Police, fire and ambulance personnel have Level C PPE, gas masks and auto-injectors of drugs to treat chemical exposure. Hospital emergency departments have yellow lines already painted on the ground to speed the set up of the victim decontamination line. All the injured have to take a shower before entering the hospital to keep it clean. This is a lesson that was learned from the Tokyo subway attack when victims streamed into the local hospital emergency rooms and closed them down. Also physicians in gas masks cannot work effectively, so patients have to be clean before they enter the hospital. Every year the Homefront Command trains hospitals for treating chemical exposure patients. One hospital in Jerusalem that has received special training for chemical patients holds an annual exercise for 300 people.

Another challenge to hospitals is the contamination of the ambient air. If hundreds of contaminated victims self-transport to hospitals they will create an airborne hazard while they are waiting to shower. Hospitals must have a plan to shut down their external ventilation intakes as long as there is a concentration of contaminated victims in the area. Even after the victims are washed down they can constitute an ongoing hazard as they exhale material from their lungs. This problem was recognized at the 2002 Moscow Theater terrorist siege, which was ended by the introduction of a narcotic gas into the theater. Even though the victims had been decontaminated inside the theater they continued to exhale and secrete material.

One way to meet this challenge is to plan for medical care to be delivered outside of the hospital building, to protect its assets, staff and patients. This technique would also be useful in the event of a contagious disease outbreak, whether related to a terror attack or from natural causes. Also, pandemic flu with contagion would be handled this was. Medical staff members look at managing outside of the hospital. Medical staff should plan for PPE, treatment and definitive care in an external facility.

One solution is to find an alternative building in the vicinity of the hospital. One Israeli plan is to use a school that is located close to the hospital for treating the minor injuries and the psychological cases. The Homefront Command would be responsible to transfer patients from the treatment site to a more appropriate hospital location, which would be in an isolated area. In ordinary cases the patients are triaged and treated at the school, and the Homefront Command physician determines which available medical facilities will receive disaster victims.

The school facility is staff using doctors from a local neighborhood medical center for immediate care of the minor injuries. The badly wounded still must go to hospital care. Because hospitals have the external decontamination lines established, the showers are already available, and if the patients pose a threat the parking lot can be used as an emergency room. During an active war a Homefront Command battalion goes into hospitals to support their capabilities.

While day-to-day responsibility for public safety rests with the police department, the Homefront Command takes over if needed. The police can close the area, and support the medics to remove injured victims. They have special training and equipment to clean the impact area so life goes back to normal quickly, which is an important element in mitigating the psychological impact of a disaster. The Israeli philosophy is that the responsibility should be held by the organization that has the advantage in training and equipment. For example, if the terror attack causes a building to collapse preventing the police from performing rescues, the Homefront Command would take over, using their more sophisticated search and rescue equipment. The police and Homefront Command train together for these missions. Because they understand that the target is saving lives, they do not fight over turf and work together.

Another mechanism of attack is the anthrax envelope. When a suspicious envelope is discovered, it is transported to the lab by the local authorities. The lab determines if the material in anthrax or another substance. Should the materials test positive for anthrax, the Minister of Defense takes command and control. While awaiting verification from the lab, the area where the envelopes were received is closed by the local authorities, and people are sent home to wait for further information. Trying to confine people just leads to panic, and treatment can be delayed up to 48 hours and still be effective, with a death rate less than 20% (Stanford, 2006).

Other types of biological terrorism may have no signs or indicators, unless the terrorists choose to issue a warning. Intelligence agencies may develop information on which to base action against the disease. Some events will be "quiet," without any indication. Biological agents may be either infectious or noninfectious. In either case early detection is important. The Minister of Defense declares a biological event based on information from a laboratory, an emergency department, a clinic, veterinarian, or Environment Department.

Delivery mechanisms for biological agents may be simple or sophisticated. Suicidal terrorists could infect themselves with a contagious disease and go to crowded areas and closely interact with people as long as he is ambulatory but febrile and start the spread of disease. The water system is postulated as a delivery mechanism for some types of biological agent. Although drinking water is carefully treated, once it leaves the plant it is vulnerable to infection at several key points in the system. However, once the contamination is discovered it is relatively easy to clean out the system and provide bottled water to the population until the water is certified again.

Israel has a surveillance system for symptoms like high fever, respiratory distress and other indicators of infectious disease, which is based in one computer system. The system's algorithm may recognize that an attack has occurred. An attack in an open place will affect people in transit, and the material may be found anywhere afterwards, carried away on their feet, clothing and even in their lungs. Israel has local and national command and control systems for managing a biological attack. The suspicion of an attack will lead to epidemiology inquiries to the Center for Infectious Diseases and clinical laboratories. A positive finding leads to the activation of the epidemic management team, and a declaration of emergency by the Minister of Defense.

Preparedness for a biological attack includes a stockpile of medical and pharmaceutical supplies. The size of the stockpile is intelligence based. There are adequate antibiotics and needles for the whole population. The real challenge is knowing when to deliver the stockpile to the victims. It may be several days between the attack and the onset of the first symptoms. Depending on the victim's health the time delay available for effective treatment will vary. The very young, the very old and people who are immune-compromised may develop the disease more rapidly than the general population, and medications may be released too late to be effective for them.

Although not as likely, a radiological dispersal device could also be used in a terror attack. Israel follows the same procedures as for chemical events. Victims are washed down to remove as much material as possible, and then sent to the hospital. The contaminated area is cleaned off with water, and the water is disposed of into a large body of water.

The Israelis have done tests dumping surrogate materials into a reservoir. They have deduced that it would take a very large quantity of radiological material to have any effect on the water, even though radiological material cannot be neutralized like a chemical or biological agent, and it degrades over a very long time span. While water disposal is not ideal, the risk seems to match the cost/benefit analysis because a radiological event is not as likely as a chemical event. While chemicals are ubiquitous and easy to transport to a mall or crowded area, radiological material is not as readily available. It is also dangerous to the transporter. Given the low likelihood of a radiological event it does not make sense to spend a lot of money on planning for the long-term recovery issues, such as disposal of the wash water.

The main impact of radiological material is psychological. People do not understand energetic materials, and presume that all types at all dose levels will produce illness and death. In fact radiological materials such as plutonium can be eaten without concern, because the body will not absorb it, but it will do damage if inhaled. One crucial way for combating a radiological attack is to inform the public about the facts of exposure, and how to protect themselves. The media, the Internet and public dialog are the best weapons against the psychological impacts attendant to radiological events.

4.9.3. PHASED RESPONSE TO A TERROR ATTACK

A terrorist attack triggers the following phased response: (a) preparation (training and obtaining appropriate equipment), (b) the first response to the incident, (c) immediate response, (d) complementary response, and finally (e) recovery. The goal is saving lives, so time is critical, with a focus of the plan being treating casualties as soon as possible. To ensure speed and effectiveness, resources are aligned to build synergy. Each organization has its own expertise. Police, fire, and medics are not duplicated but do their own work, and the Homefront Command is added.

The principles of handling an event are based on a structure. Once a declaration of emergency is made based (e.g. due to a chemical attack), reporting centers are opened to collect data for situation assessment. This data drives the decision to summon the Homefront Command and other elements of the IDF, as needed. At the same time a control room (emergency operations center) is opened to receive the situation analysis from the reporting centers and begin a formal response. Resources are called up to response to the scene, and population protection measures such as evacuation are put in place.

Mission commands are activated, and functions undertaken include victim search and rescue, victim treatment, and public information issued to the media. Information centers are opened near hospitals to help families find loved ones. The dead are photographed and put on computer, and a psychologist sits with families as they look at pictures, assisting in the identification of the victim. This spares them the sight and smell of the dead, and also allows many people to view the photos at the same time.

At the same time, organizational activity is undertaken to support the field response. First a national report is made, and then arrangements are made for salvage, isolation of the contaminated area, and decontamination to end the threat to the community. The police undertake investigation of the event, often with the assistance of the Homefront Command.

4.10. NATO and Consequence Management: The Euro-Atlantic Disaster Response Coordination Centre[17]

4.10.1. INTRODUCTION

NATO has developed a disaster response capability, with emphasis on its role in consequence management following acts of terrorism with chemical, biological, radiological, or nuclear agents. "The Euro-Atlantic Disaster Response Centre has developed principles and procedures for use in response across the alliance, as well as training and exercises" (Bretschneider, 2006).

NATO is a security-oriented organization that was not created for disaster response. Nevertheless, international disaster relief was recognized as one of the areas where enhanced practical cooperation under the umbrella of NATO could add value. Therefore, NATO Allies and Partners agreed to establish the Euro-Atlantic Disaster Response Coordination Centre (EADRCC) in June 1998 at NATO Headquarters. The concept was based on a Russian proposal in the framework of Partnership for Peace.

4.10.2. THE ROLE OF NATO AND EADRCC

The United Nations retains the primary role in the coordination of disaster relief operations. Therefore, NATO is poised to complement and provide additional support to the UN role in the Euro-Atlantic Partnership Council (EAPC) area, and not to duplicate the UN. What NATO through the EADRCC does in this area ties, of course, into a framework of international organizations such as the UN Office for the Coordination of Humanitarian Affairs (OCHA), International Atomic Energy Agency (IAEA), Organization for the Prohibition of Chemical Weapons (OPCW), WHO and the EU.

The Centre's main role within the EAPC geographical region, which includes 26 NATO member Allies, 20 Partners for Peace members, plus Afghanistan, is to coordinate, in close consultation with the UNOCHA, the response of EAPC nations to a disaster occurring within this geographical area. The role is coordination rather than direction. After 9/11, the Centre's mandate has been extended to deal also with Consequence Management after a terrorist attack in the same way that they deal with natural and technological disasters. Partner nations are equal stakeholders in all activities of the EADRCC, and all assistance is voluntary. In case of a disaster requiring international assistance, it will always remain for individual nations to

---

[17]Material regarding EADRCC is based on a presentation by Günter Bretschneider, head of EADRCC, delivered at the NATO STS-CNAD Workshop, Portugal, 2006.

decide whether to provide assistance or not. There is no commitment to react automatically. The Centre maintains a 24/7-duty officer system.

4.10.3. THE INTERNATIONAL STAFF: OPS DIVISION ORGANIZATION

The EADRCC is part of the Operations Directorate, with a reporting line to the Planning Directorate, which is responsible for Civil Emergency Planning Policy. The EADRCC has two functional areas, Operations/Planning and Policy, and Exercise and Training. There are eight staff positions in total, of which five are filled currently. The staff comes from two sections of NATO: three from NATO's International Staff (IS) (two officers and one assistant), and two Voluntary National Contribution (VNC) staff members, including a permanent coordinating officer sponsored by Switzerland. A permanent liaison officer from UNOCHA Geneva and three VNC positions are vacant.

The following are the fundamental principles of the Euro-Atlantic Disaster Response Capability.

- First, the responsibility for effective disaster response lies with the stricken nation itself, a principle adhered to by all international organizations working in the area of disaster response.
- Second, the primary role for international disaster response is with the UN, more precisely with the Office for the Coordination of Humanitarian Affairs (UNOCHA) and other specialized agencies in the UN family, such as UN High Commission on Refugees (UNHCR), World Food Program (WFP) and more importantly of course, WHO, IAEA and OPCW.
- Third, the framework of EAPC provides a gateway to civil emergency planning (CEP) capabilities in 46 nations at their origin, and thereby to solidarity to help each other in case of a disaster.
- Finally, the EADRCC was given a standing mandate to respond to a request for assistance from a stricken nation without delay and without getting further approval from Council.

However, to make this quite clear there have to be two prerequisites for the Centre's unrestricted involvement:

- some sort of disaster or act of terrorism must have occurred, and
- a stricken nation (or a UN agency on their behalf) must have requested international assistance from NATO. The assistance is always offered as a bilateral action, with one nation helping another nation, unless NATO as an organization acts collectively.

EADRCC has established Single Points of Contact in each EAPC nation, with all communication between the Centre and the nations initially taking place through these contacts.

One of the challenges for EADRCC is that many nations are net donors, such as the USA. When Hurricane Katrina struck and the USA received aid, there were no mechanisms in place to receive the goods from the international community.

4.10.4. STEPS TAKEN AFTER A REQUEST FOR ASSISTANCE

In order to react quickly EADRCC needs the following information:

- Type of disaster, when and where, population affected, area affected
- Qualitative and quantitative description of required assistance
- Timing and recommended points of entry
- National response

After verifying the request and consultations with UNOCHA, we distribute the request to all EAPC nations, UN bodies as appropriate and the EU. NATO military authorities are also on the distribution list.

In Consequence Management a mechanism has been established that allows the EADRCC to involve NATO Military Authorities at any stage of the coordination process. Essentially the EADRCC can request NATO military assistance when it comes to the conclusion that civil capabilities will not suffice to cope with a given situation. This mechanism is part of the OPLAN ABLE GUARDIAN, NATO's plan for the fight against terrorism. When having to deal with the consequences of a terrorist attack the EADRCC most likely will request NATO Military Authorities to look into the possibility to complement civil response resources with military ones.

In the EADRCC message they ask donor nations to get in contact with the stricken nation and make agreements for the delivery of assistance with an information copy to the Centre. Further on they act as a clearinghouse for information to mobilize and facilitate international assistance for the stricken nation. EADRCC distributes daily Situation Reports and updates outstanding requirements. Some nations could assist but need transportation for the assistance, so EADRCC facilitates and organizes this transportation, if requested. An operation like this will continue for 2–4 weeks under normal circumstances.

EADRCC responds to a variety of disasters, Natural disasters include earthquakes, floods, forest fires, storms and landslides. Technological hazards include chemical and radiological industrial accidents, transportation accidents involving HAZMAT, whether by air, land or water, oil spills and unexploded ordinance.

Terrorism includes attacks against critical infrastructure, industrial facilities, transportation facilities, transportation modes, chemical, radiological or biological releases, and attacks against soft targets.

4.10.5. INVENTORY OF NATIONAL CHEMICAL, BIOLOGICAL AND RADIOLOGICAL CAPABILITIES

EADRCC places its emphasis on the immediate needs of a civilian population in a catastrophic situation after a chemical, biological, radiological or nuclear incident that may or may not be the result of terrorist activity. The Inventory comprises information on nations' pertinent capabilities, which might be made available on a voluntary basis to respond to a CBRN attack against civilian populations. At the moment the inventory contains material in 10 categories, such as communications, PPE, detection, decontamination, and medical resources. For example, medical capabilities include ICU beds, field hospitals, and a variety of vaccines and pharmaceuticals.

Currently, 38 of the 46 EAPC nations are contributing tangible capabilities to this Inventory. For each category the Inventory questionnaire puts forward a definition of the capability in question, contact point information, some straightforward questions that are indicative in describing the capability, and a question on the size and readiness of the capability. EADRCC is adding more categories to the inventory, such as categories on capabilities in air transport, aero-medical evacuation, inland and ocean shipping. The revised questionnaire retains proven features of the previous version, such as simple "yes" or "no" questions, but improves flexibility with the addition of comment fields. The software tool builds on a free and publicly available platform, the ACROBAT Reader.

Since the data are sensitive in nature, EADRCC does not disseminate inventory information outside the Centre that would allow identifying certain capabilities in a particular country. The key words here are:

- The voluntary aspect of assistance
- Response without delay (24–48 hours)
- Civilian populations focus
- Mostly nonmilitary assets

The Centre has started a process to vet inventory categories against a number of scenarios in order to identify the capabilities that are likely to be needed in consequence management, and conversely to eventually remove capabilities that are not needed.

Owing to the sensitive nature of consequence management following a terrorist attack EADRCC anticipates a need for secure communication channels, which are not readily available.

EADRCC has been involved in response to sixteen major disasters between 1998 and 2006, including events in the former Soviet Union, Algeria, the USA, and Pakistan. Disaster response coordination has been in high demand lately. The Centre has been involved in a total of 10 such operations in 2005 and the 2nd one in 2006 by early March. This series started with the tsunami in Southeast Asia in the first quarter of 2005, followed by widespread torrential rain over Southeast Europe, which caused serious flooding in several countries between April and September of 2005. Altogether EADRCC has coordinated 7 requests for flood relief assistance: two requests each from Georgia, Romania, Bulgaria and one such request from the Kyrgyz Republic, followed by response to hurricane Katrina and the earthquake in Pakistan.

One of the challenges to providing assistance in a civilian environment is the issue of what goods can cross borders. There is currently no international agreement on how to cross a border and what can be taken take with you. EADRCC is hoping to develop a nonbinding agreement that would ease the passage of essential goods to and through nations, especially when road convoys of goods have to pass through other nations on their way to the disaster area. There is a lack of internationally agreed standards for various types of technical equipment. For example, in preparing for the Olympics, the Greeks wanted a stockpile of HAZMAT suits standing by, but they could not match the offer of suits to their requirements because the suits were not like their standard. Some of the suits offered were military and some were civilian. In the end NATO provided some samples that were taken to Athens for them to test and decide if they would work. In another example, the medical sector is heavily regulated at the national level, so medical goods and personnel may not be useful on a bi-lateral basis.

In 2003 EADRCC had proactive involvement in Turkey and Greece, helping to strengthen local authorities against possible problems. Turkey's request was based on intelligence that led to concerns about potential attacks. Greece received assistance in developing security for the Olympic Games. During July, August, and September three different nations provided assistance by remaining on ready status in their home country. Through advanced planning Greece and the donor nations established thresholds for when they would need assistance.

4.10.6. EXAMPLES OF NATO RESPONSE CAPABILITIES

NATO assistance after Hurricane Katrina, USA

Countries in Southeast Europe were still recovering from the worst floods in a century, when one of the most powerful hurricanes in history was about to make landfall in the USA. The EADRCC had contacted US authorities as soon as it became clear that Hurricane Katrina was heading for the American continent. An official request for assistance from the USA was received on September 3 and distributed without delay to EAPC capitals. At the US request, an EADRCC liaison officer was deployed and started working with the Federal Emergency Management Agency in Washington, DC, on September 4.

The first two offers of assistance were received on Sunday, September 4. Overall the turnout was very impressive, with 39 EAPC nations offering assistance. Offers were made on a bilateral basis, through the EU Community Civil Protection Mechanism and through the EADRCC. Following a US request, Council on September 9 authorized a NATO transport operation to help move urgently needed relief items from Europe to the USA. At that time, 38 Allied and Partner nations had already made offers for assistance. The EADRCC swiftly assumed the lead in the coordination process of the NATO operation and accommodated liaison officers from both SHAPE and Joint Command Lisbon in its structure. Coordination between the EADRCC and NATO military authorities was smooth and efficient.

The USA's accepted donations were consolidated in Ramstein Air Force Base, Germany, which served as a strategic air hub for onward transportation. Donations were either moved by road or by NATO Response Force tactical airlift under the command of Joint Command Lisbon. All cargo consolidation from European donating nations had been completed on September 19, 2005. More than 90 flight hours were flown by French, German, Greek and Italian C-130 and C-160 tactical NRF transport aircraft.

The first NATO relief flight arrived in Little Rock, Arkansas on September 12 with 12 tons of donations from the Czech Republic. The last NATO cargo flight, one Turkish C-130, arrived on October 2 in Little Rock with donations from the Czech Republic, Romania and Slovakia. From the beginning through October 2, twelve NATO cargo flights had taken place using aircraft from NATO's Airborne Early Warning fleet (three B-707s), Canada (one Airbus A 310), Ukraine (one AN 124) and Turkey (two C-130s). Altogether, almost 200 tons of relief goods had been delivered through the NATO air-bridge. The EADRCC operation was suspended with the distribution of a final report on October 2, 2005. It should be noted that this was the first time that NATO activated elements of the newly established NATO Response Force (NRF) in a purely humanitarian role.

NATO Pakistan Earthquake Relief Operation

Two days after a devastating earthquake had struck the country, Pakistan authorities, on October 10, approached NATO with an official request for assistance. Within one day, on October 11, Council decided to put a strategic air-bridge to Pakistan in place to move urgently required relief items. Again, as in the Katrina relief operations, elements of the NRF had been activated to support a humanitarian relief operation. The EADRCC, working in conjunction with NATO military authorities, UNOCHA, the EU and Pakistan authorities was tasked to coordinate all offers from NATO and Partner nations that had also requested NATO transportation assistance. The first NATO relief flight to Pakistan had left on October 12 and arrived on October 14 carrying donations from Slovenia. The NATO relief operation in Pakistan came to an end on February 1, 2006. The last flight of the air-bridge arrived in Pakistan on February 8, 2006.

From the beginning, NATO's air-bridge has been moving urgently needed tents and blankets to Pakistan, with the majority of the relief items being provided by UNHCR. Priority items were shelter, medical supplies and warm clothing. In the end, 42 EAPC nations have been providing assistance to Pakistan, either on a bilateral basis, through the EU Community Civil Protection Mechanism or through the EADRCC. Financial donations and donations in kind to the NATO air-bridge allowed EADRCC to charter commercial airlift to augment the fleet of NRF tactical aircraft being employed. Since the first NATO flight from Slovenia, more than 160 relief flights delivering almost 3,500 tons of relief goods have taken place. These included tents, sleeping bags, heaters, mattresses, blankets and many tons of medical supplies. The NATO air-bridge has been used by 19 EAPC and 2 non-EAPC nations and, of course by UNHCR, WFP and UNOCHA.

The Council also authorized, on October 21, deployment of a NATO Disaster Relief Team to Pakistan for 90 days. This team was comprised of NRF and other national military and civil elements under the command of a NATO force commander. The capabilities deployed were a multinational field hospital led by the Netherlands, helicopters from Germany (4) and Luxembourg (1), three light engineer units (companies) from Spain, Poland, and UK, one heavy engineer company from Italy, and a helicopter re-fuelling station from France. The size of the force was about 1,000 personnel.

Overall the operation provided impressive international assistance. More than 8,000 patients were treated, and clean water was produced to care for 1,000 people per day. Over 41,000 cubic meters of debris was cleared from roads and the roads were repaired. 110 schools and shelters were reconstructed. Upon NATO's withdrawal on February 1, 2006, medical supplies and equipment were handed over to local authorities, while the helicopters and the French fuel farm continued to operate under bilateral arrangements.

4.10.7. TRAINING AND EXERCISES

The EADRCC has an ambitious exercise program with one large field and one tabletop exercise each year. The regional approach is used, and exercises are coordinated with the UN whenever possible. Six major field exercises have been conducted so far. The first field exercise took place in Ukraine in 2000. The exercise, called "Transcarpatia," used a flood scenario.

In 2002, the EADRCC conducted two major exercises in cooperation with the host nations. The first one was held in Croatia in late spring with the scenario of "Wild Fires." The second one, called "Bogorodsk 2002," which took place in Russia in September, used a terrorist attack on a chemical facility as the scenario.

Again in 2003, two exercises were conducted, "Ferghana 2003" in Uzbekistan in April/May, with floods and landslides and a chemical facility accident as a scenario. In October, "Dacia 2003" was conducted in Romania with a dirty bomb scenario.

The recent consequence management exercise "Joint Assistance" in October 2005 was jointly organized with OPCW and hosted by Ukraine. The exercise was built around a scenario of a terrorist attack against civilian populations using chemical weapons. Fifteen nations participated in various exercises, with the largest group from the host nation, and 200–300 participants from the international community. Interoperability capabilities were tested using a regional model. The exercises did include UN involvement, but the UN does not do CBRN, only natural disasters. Disease outbreaks can be exercised including WHO.

4.10.8. NATO SHAPE/ALLIED COMMAND OPERATIONS[18]

The Supreme Headquarters, Allied Powers Europe (SHAPE) is also concerned with consequence management after a terrorist attack. NATO needs to be changed to face a variety of new threats using fourth-generation warfare. This requires a change of mind-set from war fighting to problem solving. Strategic commands are organized for transformation, while Operations

---

[18]Material regarding SHAPE is based on a presentation by Alan Butterfield, delivered at the NATO STS-CNAD Workshop, Portugal, 2006.

includes new capabilities and concepts, integrating maritime, land and air (Butterfield, 2006).

In the recent past NATO's response was static, facing east. Now its response force must be able to deploy in a 360-degree way, and up to 25,000 must be rapidly deployed. This model was tested in the real world responses to Katrina and Pakistan within just a few weeks of each other. NATO includes a multinational Joint Assessment Team that was activated following the 9/11 attacks on the USA under Article V "Collective Defense." The World Trade Center attacks killed citizens of more than 60 nations.

The Global War on Terror involves 26 member states. Article V was not invoked for either London or Madrid's transportation bombings. The military response was mounted through NATO member states with UN cooperation. Training for the war against terror is based on building national capacity. NATO has funded a Center of Excellence in the War against Terrorism based in Turkey.

There are also terrorism related activities carried out at the brigade level (5,000 troops). For example, as was mentioned during the EADRCC presentation, it took only 4 days to establish the air bridge to Pakistan for relief. The second level response was the field hospital and mobile medical capabilities, helicopters, engineering units, water purification, and refueling farm. NATO and SHAPE are considering getting prior agreement from the political level for future deployments in specified circumstances to speed action.

The strategic elements of the military are developing the concept of operations and plans for defense against terrorism, including consequence management and disaster relief. This activity has spread far from the NATO area. For example, NATO conducted Mediterranean maritime operations to protect high value targets against piracy and terrorism as they pass through the Straits of Gibraltar, as part of the Article V operation against terrorism. In addition the Joint Information and Analysis Center is considering developing a fusion cell for intelligence from various sources within member states. A more sophisticated intelligence center would be useful for high visibility events. They would also contribute to the evaluation of the vulnerability of critical infrastructure components due to natural disasters.

More assistance might be offered, but the cost principle limits the response of members. It is always the same nations being asked to contribute, so some limit to what they are willing to give. Alternatives to encourage more participation are not readily obvious. "Costs lie where they fall" (Butterfield, 2006), and nations with smaller economies may find contributions too hard on their budgets.

4.11. European Union Civil Protection Mechanism[19]

4.11.1. MAINTAINING PREPAREDNESS

The EU Civil Protection Mechanism was established on October 23, 2001. Before the 9/11 attacks the EU's response to civil disruption was limited to mitigating the effects of natural disasters. Their capabilities now include CBRN. The 30 EU states have internal response and preparedness capabilities. In general the domestic departments of Justice and Home Affairs manage the response and criminal justice elements of a terrorist event.

The EU's response is focused on monitoring events and managing an information center, using the Common Emergency Communication and Information System (European Union, 2007a). In recognition of the threats faced by member states, the EU holds joint training and exercises. For example, there were four exercises in 2005, and two in 2006 on terrorist attacks. They have an exchange of experts, an action program for civil protection, and a database of civil protection capabilities that members can draw upon (European Union, 2007c).

The Monitoring Information Center (MIC) provides technical support to the response. It provides links to other EU members for information regarding common threats, such as pandemic flu. Although the MIC is an alert center, during EURO 2004 it worked in prevention for first time because of terrorist attack scenarios (European Union, 2007a).

If there is an intervention inside the EU, the host/victim nation maintains operational control. For example, Portugal suffered serious wildland fires. It requested seven aircraft capable of dropping water on the fires. These resources were added to the resources available to the Incident Commander. The EU also participated in providing assistance to the USA during Hurricane Katrina. They helped to coordinate assistance from 21 European countries.

The EU recognizes that there are areas of community cooperation that need further study and advancement in implementation. They have allocated resources for a coordinated program of pilot programs call the Action Programme for Civil Protection. It has rolling categories that represent the highest unmet priorities of member states. The Action Programme was funded from 1999 through 2006. The eight categories listed were improving public information, using lessons learned from terrorist and other confidential actions, psychosocial after care, risk assessment and risk management,

---

[19]Material regarding the European Union includes information from a presentation by Sasa Borko at the NATO STS-CNAD Workshop, Portugal, 2006.

vulnerable populations, assessment methodologies, workshops for disaster medicine, general contingency planning, or any other topic within the Programme, and fire prevention and safety measures (European Union, 2007b).

4.11.2. LESSONS LEARNED

The EU needs to develop a regulation to provide a mechanism for funding civil disaster response. They also need to better address shortages of transport and equipment. They need better response capabilities for response to all disasters, especially terrorism. Events with cross-border effects need to be studied to create a model for response, including premade agreements regarding border crossing with equipment.

Recognizing the critical importance of cross-border response the EU allocated €5.6 million to fund five projects. "These projects are aimed at raising awareness and providing a framework for closer cooperation in civil protection in the fields of cross-border early warning, coordination, and logistical tools with a view to preventing or at least minimizing the consequences of natural disasters" (European Union, 2007d). The pilot programs are aimed at a variety of natural hazards. Flood Management Cross-Border and EU-USAR: Relief Cross-Border projects are being led by Technisches Hilfswerk, Germany. A cross-border team for flood fighting in Latvia and Lithuania is being led by the Jelgava City Council, Latvia. Fire 4 is a project of Italy, Spain, Portugal and France to fight forest fires, and is led by the Ministère de l'Intérieur et e l'Aménagement du Territoire, France. Italy, Portugal and Germany are working together on a mobile command and earthquake assessment capability, led by EUCENTRE, Italy. Finally, EU Flood Command is a joint program of the UK, Ireland and Sweden to develop a maritime search and rescue capability for coastal flooding. Each of these projects is viewed as a pilot to be spread across the EU if successful (European Union, 2007c).

The Madrid bombings showed the need for solidarity among all EU members against terror attacks. The EU Council is assessing the capability of member states, including measurement against seven generic scenarios, including terrorism, to determine what assistance they should make available for each. Once the assessment is completed, shortfalls may be addressed.

The EU military staff is creating a database with military assets available for military actions against CBRN attacks, and it is now being extended to all natural and man-made disasters. While there is some overlap with the NATO database, the EU database is used only for analytical purposes. The nations are asked each time if they want to participate in the response. The ARGUS rapid alert system will enhance the capability to react quickly in a coordinated manner.

Within the EU, the organization can be asked to find the needed assistance or goods, and the receiving state can be asked to pay for the assistance. Member states can save money by coordinating resources to focus on their greatest needs, and look to mutual aid for the seldom-used equipment. However, there may be internal political reasons why member states do not want to appear dependent on outside resources. Both NATO and EU resources can be requested by member states.

4.12. Organization for the Prevention of Chemical Weapons[20]

4.12.1. ROLES AND EXPERTISE

The Organization for the Prevention of Chemical Weapons (OPCW) has as its mission the enforcement of the Chemical Weapons Convention, and the elimination of chemical munitions from the world's offensive weapon stockpiles. 175 nations have signed the treaty, including 29 member states in the sphere of influence of the EU and NATO. The OPCW role in response to terrorism "is to offer expertise after a major terror attack" (Trapp, 2006).

The OPCW works to ensure that evidence is collected and preserved at the site of chemical weapons accidents or attacks for later use in legal proceedings. They also oversee protection and verification of the existing stockpiles of chemical weapons that are in the process of being destroyed by member states, to ensure that they are not stolen by rogue non-state terror groups. It is an intergovernmental organization that is focused on eliminating the chemical weapons of states.

OPCW has an emergency response mechanism in case someone uses a chemical weapon against a state party, whether the attacker is a state, terrorist, or chemical accident. They work to develop national measures to control toxic chemicals, and have an inspection regime to ensure compliance with the treaty.

The OPCW does not define chemical weapons by a specific list of weaponized substances. Rather it looks at chemicals with the characteristics necessary to kill. A chemical weapon is any toxic substance that is used against human beings with the intent of causing harm and death. Therefore toxic industrial chemicals (TICs) can be diverted and used as weapons.

---

[20]Material regarding the OPCW is based on a presentation by Ralf Trapp at the NATO STS-CNAD Workshop, Portugal, 2006.

4.12.2. INTERVENTION CAPABILITIES

The OPCW only intervenes in an event at the request of a state party. Response to a chemical attack must be instant, counted in minutes, if lives are to be saved. OPCW's role is not as a first responder, but to come after the initial response to provide expert advice on the cleanup and evidence preservation. They also assist with the development of protective capability against chemical weapons for planned events, such as preparation for the 2004 Olympics. Assistance is based on pledges by state parties, with a database of resources from member countries, so OPCW coordination with NATO and the EU at the requesting party level is needed.

After an attack the OPCW assists with collecting and protecting and preserving evidence for prosecution, and for evaluating the need for additional consequence assistance from neighbors. The 2003 conference recognized that chemical facilities that could be used by terrorists need special preparations for evidence collection, and to be sure that a plan is in place for a proper investigation to see who did it.

The investigation protocol is based on what was done in the 1980s investigations, but the world has changed, so some protocols may not work. The OPCW needs to update the operational side of its response to better meet the purpose of the investigation. The first investigators are the forces of the member states. Members without much local resource for HAZMAT investigation may be more dependent on OPCW response capabilities. In 2005 OPCW held an exercise with Ukraine and NATO, and OPCW's role was to identify the chemical used as a weapon.

One problem that must be addressed is the potential conflict between the criminal investigation and the ongoing emergency management. Teams should be integrated into the command system, including the chemical field analysis group. The OPCW Coordination and Assessment Team only respond based on a request from the victimized state.

One of the current gaps in response is the inability to conduct analysis of biomedical samples in the field. There needs to be a more rapid means for determining which victims are being damaged by the chemical, and which are suffering from psychological trauma. People can be in the area but remain unexposed. It is very hard to detect a chemical agent in body. Blood and bodily fluids often provide the best indicators, but collection and analysis takes time. This is not what OPCW staff members are trained to do. However, protocols for victim identification through medical analysis need to be established before there is a widespread event. One challenge is how the investigators will maintain their skill level in dealing with human victims when this is outside the norms of their daily work. Joint training with medical technicians would be one solution to the need for better victim analysis.

Because the assistance of OPCW is driven by the request of the host nation, if their capabilities are weak they may not know to ask for the right advice and the right assistance. There should also be metrics for managing these requests and getting the teams on site. Realistic scenarios should be established to test response times. In the past many of the exercises have been driven by which nations and resources wanted to participate, and to make room within the exercise for maximum participation. In order to have more realistic exercises the scenario would have to be drawn closer to a real event, and then participants would have to be modeled on reasonable expectations and time frames.

4.13. The Way Forward

NATO and EU member states and their allies are challenged by the terrorist threat. Many nations are using an all-hazards approach to maximize the benefits of the expense of preparing for terrorism. Common across the Alliance is the recognition of the need for a strong and well-understood chain of command. Further, with the transnational character of terrorism, it is possible that mutual aid across borders may be required for an adequate response to save lives. This makes the need for a common understanding of command and control crucial.

Mutual aid has been coordinated through EADRCC, but the experience of Hurricane Katrina showed the areas that need additional strengthening. First, there should be consideration of multinational funding for the Alliance's response to civil events. Second, there should be consideration of how response to terrorism situations would differ from response to civil disasters. Would the costs still "lie where they fall," making it difficult for the smaller nations to contribute needed resources? Would attacks with weapons of mass destruction/disruption/killing change the NATO response? There was NATO assistance in "providing AWACS command-and-control aircraft to help defend US airspace in the days after the 9/11 terrorist attacks," flying air cover for the USA (Herbert, 2003), but no response to the train and subway bombings in Madrid and London. What standards will be used in future attacks to determine the level of NATO involvement?

Finally, the EU created open borders, which the terrorists can traverse with impunity, while maintaining all the rights of national sovereignty that make it impossible for law enforcement to pursue. Technologies like GIS and air monitoring will make it clear how far the consequences of such an attack travel on the wind, water and across the topography. Will the Alliance, working with Interpol, aim for a new standard of cross-border credentialing of law enforcement personnel, medical personnel, and engineers? After a WMD/di/K attack there will be no time for international negotiations about the law of response. These issues should be addressed in NATO now to ensure that when terrorism escalates to CBRNE, which it already has in Japan with Sarin (Olsen, 1999) and in Iraq with IEDs with chlorine cylinders (Parsons, 2007), the Alliance will be well prepared to rescue the victims and confront the enemy to prevent future attacks.

Dr. Amos Guiora, retired lieutenant colonel in the Israeli Judge Advocate General Corps, and professor at Case Western Reserve Law School, has undertaken a cost–benefit analysis of terrorism preparedness. His conclusion is twofold. First, this is not a war, it is a new state of civilization for perhaps the next 100 years. Second, nations have no choice but to use their resources to defeat the terrorists in the long term, both through effective counterterrorism operations, and effective social betterment in nations that harbor terrorists (Guiora, 2007).

The NATO states will be confronting transnational terrorism into the foreseeable future. Improving the capabilities of first responders within the Alliance is a critical task for each nation-state, and for the collective effort. The current capabilities of each member must be well integrated, and all skills must be ready for deployment in life-saving missions, across borders, with no notice. Effective planning now will ensure this crucial outcome.

5. RESEARCH AND DEVELOPMENT NEEDS IN FIELD RESPONSE TO TERRORISM

Modern liberal democracies need to develop innovative methods and equipment to deal with the new terrorism-driven urban warfare. The spectrum of response to the demand for better civil protection ranges from an increased military build up, to emphasis on research and development (R&D) in adapting off-the-shelf technology available from the military sector to reduce the vulnerability to terrorist attack. In the past such R&D was driven by market forces, but in the war against terror the safety of the citizens cannot rely on marketing or exaggeration of product capabilities. In the era of cross-border terrorism there must be a prioritized list of needed advances agreed upon by the NATO member states to ensure that scarce R&D funds are spent to achieve the greatest amount of improved protection possible.

Definition of civil protection aims is a highly political activity, but one that is not politically popular. Open liberal democracies are used to operating with strict civil rights protections for individuals, while the population expects to remain safe. Terrorists can be members of the society they are attacking, as in the London case study. Therefore, there may not be clear ethnic lines that divide terrorists from peaceable members of the society. For example, while in Israel bus drivers can be taught to recognize potential terror suspects by ethnic differences among the population, most European societies cannot rely on such overt methods of surveillance against terrorists. Government agencies and NGOs are working together to find ways to protect everyone's civil rights while also guaranteeing the public's safety. Everyone agrees that it is government's responsibility to protect the population by preventing acts of terrorism, but the only way that this can be done, within the legal the constraints of a free society, is to enhance technology for surveillance, detection and characterization of WMD/CBRNE weapons, in order to ensure that apprehensions have adequate probable cause for legal action.

The government's surveillance abilities concerning nuclear, radiological, biological and chemical risks need to be improved. Releases of harmful material may occur because of industrial uses, as well as through possible attack. Whether the dangerous material is released into the community through industrial accident or through terror attack, the ability to maintain the public's safety currently is limited. In most communities population protection is based on public warning through the media, and the technique

F. L. Edwards and F. Steinhäusler (eds.), NATO and Terrorism, 161–177.
© 2007 *Springer*.

of self-protection called "shelter-in-place." There are no stockpiles of personal protective equipment for the effected populations, and for persistent or acutely HAZMAT, shelter in place may not be enough.

A program of R&D aimed at meeting the priority needs of first responders for communications, personal protective equipment and surveillance/detection/characterization would enhance the effectiveness of the public agency financial support for R&D available to technology companies. The NATO workshop in Portugal in 2006 was focused on examining the capabilities and gaps in responding to the new urban WMD/CBRNE terrorism, and to considering priorities that could be agreed upon among the allies. Such an agreement will enhance future cross-border coordination, as well, as response capabilities develop along similar pathways.

The Heritage Foundation has concluded that multinational approaches to R&D for counterterrorism equipment developments is essential. "Countries have limited financial, human, and other resources available for homeland security. Winning the long war to disrupt transnational terrorist networks will require international collaboration in researching, developing, and sharing homeland security technologies. By facilitating greater cooperation, spreading research and development (R&D) costs, and taking advantage of synergies, the United States and its allies can extend the impact of their homeland security programs, entice businesses and entrepreneurs with larger numbers of potential customers, and take advantage of the continuing internationalization of the global market for security technologies" (Carafano et al., 2006). The USA is investing over $6 billion annually in the R&D efforts of industry; and other NATO partner states are also investing in the development of needed technologies and equipment. A prioritized research program could enhance the benefit of these funds to the alliance by avoiding duplication and encouraging collaboration.

5.1. The Catastrophic Risk Model as a Basis for R&D on Emergency Management

Planners for national and local emergency preparedness need to account for the fact that in future first responders will have to be able to operate under extreme boundary conditions, such as urban warfare. This capability of the first responders is critical for the mitigation of mass casualties and infrastructure loss within a live battle space. The contributions of research and development (R&D) to first responder operations include: (a) management of the CBRNE weapons' effects; and (b) assessment of options for overwhelming or mitigating terrorist strategies and tactics.

Three tiers of research activity are recommended in a *catastrophic risk model* to provide a hardening of vulnerabilities and a capability to overwhelm terrorist activity in the immediate, medium and longer terms.

Table 1, "CBRNE^2 Risk" and Table 2, "Three Tier CBRNE^2 Science Innovation and Systems Acquisition to 2025" provide a modeled approach to research. This multidimensional model covers the decision-related parameters: risk; time; consequences; and countermeasures. Thereby it provides a basis for quantification of research. By relating research to vulnerability, and providing a comparison of research applications across the different catastrophic risks, it is possible for governments to determine acquisitions from national laboratory programs, smaller innovative groups and commercial defense suppliers. Additionally, the model makes it possible to plan efficiencies with multinational contributions toward any significant program to overwhelm terrorism on a global scale.

Within the model, the risk assessment is based on current trends toward urban warfare in the Middle East (Iraq, Saudi Arabia, Jordan), Russia, Singapore, the USA, the UK, Spain, Turkey, and Germany. As an example, the London bombings of July 2005 were shock wave attacks against urban infrastructure by small networks of four or more persons. Since the July bombings there have been at least three significant plots for further attacks that have been disrupted. The Madrid rail bombings in Spain in March 2004 were also followed by threats that were disrupted. Similarly the siege at the school in Beslan (Russia) of September 2004 was a shock attack and subsequent threats have been disrupted. Continuity of attacks and shock waves with the goals of overwhelming economic infrastructure, creating public fear, and destabilizing governments are global terror strategies.

A quantified risk assessment across all three tiers of the model in Table 2 is based on catastrophic risk, where threats are against critical functions with irrecoverable consequences. This model is an example of risk modeling for research and acquisition of research programs. This level of analysis takes the scientific research and development beyond an historical focus on major incidents and emergency preparedness.

While countries moved quickly to create budgets and significant research programs after 9/11 there are many changes that may be required because of the rapid development of capability by terrorist groups. The changes in terrorist capability may be significant within short time periods of 6 months or more. The growth in terrorist capability requires marked changes in the level of protection that first responders, modeled on traditional contingency plans, can provide in cities.

Additionally there has not been a rationale for applying defense materials and technology for counterterrorist protection across the range of CBRNE risks. The rationalization of the use of the defense sector to provide for national resilience requires encouragement from a formal resilience acquisitions policy. Modification of sophisticated defense technology into systems for the protection of urban infrastructure and the public, and for global network-enabled counterterrorist systems, requires a quantified catastrophic risk analysis.

5.2. R&D for Urban Warfare and Terrorist Capability

In the following section a profile of urban warfare is established, based on the international experience gained over the last 5 years.[1] This profile is quite different from the level of expectations in most European cities and in most other countries.

Coordinated logistics and attacks: The attacks are not single attacks that are easily managed but may involve a range of attacks on different targets in any day, an overwhelming use of one form of weapon, or the application of different weapons to targets; and attacks are sustained every day rather than being controlled by the security forces. Across the urban environment terrorists can rely on an extensive distribution of weapons, training material and motivational material. There is a manufacturing capability to provide improvised explosive vehicle-borne devices from within the city, and possibly from within other cities nearby. Within the urban environment there may be a close proximity between the manufacturing site in a garage and the deployment of the devices by suicide bombers, or by terrorists who place the vehicle and use command wire or mobile phone initiation to selectively attack convoys of the security forces.

Suicide bombers: Terrorists are likely to have the capacity to deliver over a hundred vehicle-borne improvised explosive devices (VBIED) in a month. Fifty of these VBIED's could be suicide delivered. In addition, in any month there may be 10 or more attacks by suicide bombers using concealed improvised explosive devices. The use of suicide bombers will probably be a consistent strategy over several years, with terrorists showing the capability to arrange 60 or more persons to be suicide bombers in any month. As the majority of these attackers appear to come into the targeted country as recruits from elsewhere, the potential for concerted waves of

---

[1] The raw data for the analysis in this section is drawn from the Hazard Management Solutions (2001-2006) Triton Report, www.hazmansol.com, United Kingdom. Unpublished data base tool of terrorist attacks each month across all countries. The gross database content covers over 33,000 incidents since 2001 and the current monthly Triton Report data from Iraq covers over a thousand incidents of note.

attacks with suicide bombers in any European country, the USA or any other country is a factor to take into account in catastrophic risk planning.

Civilian targets: The civilian population is directly targeted in mass casualty bombings from vehicle-borne improvised explosive devices (VBIED), resulting in deaths numbering in the hundreds. A coordination of suicide bombers with one or more VBIEDs may mean that a terrorist wearing a suicide belt will move alongside the responding emergency services and civilian rescuers to detonate amongst the crowd. Another suicide bomber may attend the hospital accident and emergency reception area alongside the survivors and detonate a device. In assassinations, the targeting moves between political leaders, religious leaders, leaders in the police and security forces, and officials in key ministries.

Weapons and tactics: The range of weapons may include one or more SAMs deployed against aircraft, RPG's, rockets and improvised rockets, mortars and improvised mortars. Generic command initiated improvised explosive devices may constitute 40% of the thousand or more attacks in a month. Governments will face a consistent development of terror tactics, with terrorists becoming more adapted to the targeting of individuals, defense, police and security services, critical infrastructure and government. Weapons will be positioned more effectively and the means of breaching "hard" or well-defended targets will be identified and practiced. The ability to overcome countermeasures will be identified, both in the ability to use higher energy weapons against armored vehicles, and to position devices closer to targets through subterfuge or from assessment of vulnerabilities. This will include, for example, the use of police uniforms or police cars, and the delivery of suicide bombs by persons within a training course; or the assessment of electronic countermeasures against improvised explosive devices will lead to devices that are initiated through water pressure with no electronic signature. Overwhelming force is a tactic by terrorists and this threat is well beyond isolated attacks that have been used in countries such as the USA, the UK, or Spain. Groups up to one hundred may join up to form attack groups armed with conventional weapons in several vehicles, thereby presenting significant firepower. The use of rockets and missiles will be a threat to be reckoned with, for example pilots of emergency crews will face instances of being fired upon by ground fire with various weapons, thereby creating a risk to air space. There is likely to be some losses of aircraft providing emergency services from various forms of attack. Significant bombings of key government targets (e.g. ministries, embassies) will be used by terrorists to show the potential of striking at the heart of government.

5.3. Strategic Operations Model for Scientific Innovation

Catastrophic risk is the top end of the risk dimension. It is a level of risk that potentially overwhelms the capabilities of systems to the extent that an impact will cause irrecoverable damage. Attacks that focus on the center of government, economic stability, critical infrastructure and mass casualties are risks at the catastrophic level. A summary of examples of catastrophic risks from CBRNE terrorist attacks is outlined in Table 3.

The design of research into countermeasures to protect life and critical infrastructure vulnerabilities against urban warfare can be based on a model of catastrophic risk. The problem for governments and first responders is how catastrophic risk is to be assessed, planned for and managed to ensure the survival of the society.

Catastrophic risk goes beyond the general major incident planning that has been the basis for emergency preparedness in NATO member states. Since 9/11 many countries have increased commitments of resources and training, but catastrophic risk requires a new set of assumptions and risk assessments. An increase of resources for a contingency plan does not necessarily protect against a total loss of inventory of a nuclear reactor facility, or the loss of all of a country's elected politicians in a biological attack on a building.

A strategic operations model (see Table 3) for scientific innovation has been developed to match research and innovation to CBRNE risk that is outlined in Table 4. Table 4 identifies scientific solutions as counter-measures that may reduce the risks identified in Table 3. A three-tiered timeline in Table 4 covers 20 years of research and development, and shows examples of solutions that are relevant to mitigating critical national vulnerabilities.

Failures at critical points are a basis for assessment of where the society may fail under a real attack and what components of the society cannot be accepted as loss events, i.e. critical points where resources must be committed to protect the fabric of the society against a catastrophic level of risk.

At the catastrophic level, risk assessment focuses on the key indicators of failure based on an analysis of what controls or measures exist for protection, which of these will really work in an attack, and whether the controls will protect critical functions under the conditions of a real terrorist attack. Although many counterterrorist plans and measures may exist, the effectiveness of these countermeasures needs to be tested against weapons' effects, terrorists' tactics and strategies, as well as the real operational capability of responders and the capability of the public to survive.

TABLE 3. CBRNE² Risk

Catastrophic Risk CBRNE²	Explosives	Chem/Bio	Nuclear/ Radiological	Electronic	Psyops	OpeartionsRisk Summation
Terrorist Attacks	Large VBIED < 2 ton explosives	Hydrogen Cyanide	Explosives/ Dirty Bomb Nuclear	Air traffic Rail Shipping	Global news incidents on explosions/ hostage taking	+/- for Operations Outcomes based calculation of
VBIED—vehicle borne improvised explosive device				Financial		operation moving risk down toward 0 or failing to
VBIC—vehicle-borne improvised chemical device	Attacks within confined space	Chlorine Anthrax/Avian	IND - 1.5-15kt Tactical	Water/sewerage		impact so that the risk is open
IND—improvised nuclear device	< 20 kg explosives Tunnels,	Flu Ricin	Weapons VBIND	Power supply and distribution	Web videos and chat rooms	and represented as a 1.
VBIND—vehicle-borne improvised nuclear device CBRNE²—chemical/	buildings, industrial plant	VBICD	Nuclear weapon on Nuclear Facility	Mass media First responder Communications Defense	Videos	
biological/radiological/nuclear/ extreme explosives, electronic/electromagnetic	Fuel/air explosions through release of fuel or chemicals from	Storage/transport Release of chemicals	Nuclear weapon or radiological materials released on	Communications	Trigger events through media for	
Psyops—psychological operations	sites/transportation	Nerve Agents— persistent and	Ground Air		Recruitment Finance	1. Overwhelm terrorist capability with preemptive
Chem/Bio—chemical and biological	Shaped charge munitions	non persistent	Ship-borne Rail Underground Hidden Source Nuclear Facility		Fear and Power CBR/phobia Shock attacks Shock wave Sustained	measures 2. Prevent and interdict some

Catastrophic Risk CBRNE[2]	Explosives	Chem/Bio	Nuclear/Radiological	Electronic	Psyops	OpeartionsRisk Summation
			release of inventory		attacks	attacks 3, Respond and mitigate 4. Recover critical functions
					Coordinated attacks in-country and international	

TABLE 4. Catastrophic risk model: three-tier CBRNE[2] science innovation and systems acquisition to 2025

Catastrophic risk CBRNE[2]	Explosives	Chem/Bio	Nuclear/Radiological	Electronic	Psyops	Opeartions Risk Summation
National critical function vulnerabilities examples	Political & government Parliament and government buildings International trade and transport including shipping	Mass casualties	Energy availability	Financial transactions	Fear of travel Panic retail Recruitment	Operations: Set Outcomes to protect from or mitigate critical functions loss

Catastrophic risk CBRNE[2]	Explosives	Chem/Bio	Nuclear/Radiological	Electronic	Psyops	Operations Risk Summation
TIME 3 tiers for research and acquisition	TIER 1 Immediate 0–3 years		Tier 2 medium to 10 years		Tier 3 to 2025	Post 2025
Science innovation examples	Preemptive small CBRN capable defense teams in cities	Self-help capabilities of individuals, work teams, households, local and national government agencies trained and tested	Construction materials for CBRN mitigation in urban space and within buildings, transport to absorb, deflect, mitigate	Shielding SCADA legacy systems hub protection	Disruption of terror recruiting campaigns	Outcomes summary of mitigation of risk from the countermeasure
Preemptive and first responders systems	Attach industrial chemists/physicists to Fire Services first response	Community prevention programme	Conversion of defense materials for civilian infrastructure protection	Networked Global C4i system links business and intelligence on financial and patterns reports.	Community based projects	
C-T counterterrorist					Analysis of network locations and risk of recruitment	
C4i—command, control, communications, computing, intelligence	Contamination cleanup for hot spots—urban space, buildings, critical infrastructure			Global hub for alarms and c-t campaigns	Preparation of public attacks with self help confidence and power of local recognition of pattern changes from terrorist activity	
SCADA—supervisory control and data acquisitions						

Catastrophic risk CBRNE^2	Explosives	Chem/Bio	Nuclear/Radiological	Electronic	Psyops	Opeartions Risk Summation
Research Costs						
Estimates for						
cost–						
effectiveness						
analysis in risk						
mitigation						
Government						
acquisitions						
policies/budgets						
Acquisition						
decisions						
Research						
program						
decisions						

Note: The Critical Functions Risk Analysis provides combinations across all four dimensions based on risks in Table 3 and in Table 4 the Rows listing many factors for Vulnerabilities, Time, and Research Innovation as countermeasures. Analysis of factors across all these dimensions provides decisionmakers with a choice of policy options based on mitigation potential of the research, cost effectiveness and comparisons of risk reduction across some of the significant CBRNE^2 risks. This provides a powerful analytic tool for quantification of catastrophic risk for short, medium and longer-term calculation and testing of impact of science and system innovation and development. The analysis of cost benefit and cost effectiveness of proposed research programs provides a quantified system of calculation for national acquisition policy in counterterrorism.

Table 4 follows CRNE^2 Risk examples identified in Table 3: CBRNE^2 Risk

5.4. Timelines for Deliverables

The contributions of science to counterterrorist measures can be measured if the outcomes of each contribution are related to critical national vulnerabilities. Without structuring scientific development into a model, the general focus of scientific endeavor may follow random scientific applications, or be submerged by the programs delivered into the commercially weighted defense industry sector. A focused scientific approach with the use of a model, such as illustrated in Tables 3 and 4, allows a structuring of contributions over periods that are relevant to grant submissions and to the time taken for innovations to mature and be tested in the field.

Three timelines or tiers for scientific research and development are required for the planning of effective scientific research and innovation. Counterterror measures need to be planned for 5, 10, and 20 years of applications:

- Short-term solutions will be required to protect cities in Europe, the USA, and the Middle East, if clear and present dangers have been identified (e.g. warning by the intelligence services about impeding terror attacks involving a "dirty bomb").

- Mid-term solutions involve planning of networked solutions for the G8, UN and NATO and strategic alliances of countries. This should create conditions to overwhelm terrorist recruitment, planning and operations.

- Research and development activities over 20 years should cover countermeasures that can be embedded into urban space, e.g. by applying some defense materials and technology to buildings and transport. The example of protective embedding of materials and systems is based on the possibility of converting defense technology into urban applications. A long-term investment in the build of networked systems for international partnerships between nations and business would stabilize economies, using a global information system on terrorist finance and operations.

The stepwise development of government and private sector acquisitions policies can be conceptualized through a three tier system for applications across immediate and longer term time lines, with relevant benefits calculated over these periods.

It is proposed that national scientific laboratories and other focused innovative groups identify and present the case for research, and respond to national acquisitions policies based on vulnerabilities identified in a catastrophic risk analysis. This is a real deliverable from research and innovation toward overwhelming or mitigating terrorist risk.

The use of modeling for acquisitions policies by governments and the private sector would encourage scientists to think about the rapid development of technical capability within the terrorist groups and consider the real outcomes of proposed research to vulnerability reduction, i.e. the operational capacity of mitigation solutions against future trends in terrorist technology.

In Russia there has been a strong focus on organized plans for research on terrorism and technology. In November 2003 in Russia the scientific-practical conference "Problems of technological terrorism and methods to prevent terrorist threats" was held. One of the outcomes of the conference was the development of prospective directions for counterterrorism research focused on technological terrorism. This list is as follows:

- Studying the reasons for and preconditions of terrorism, and developing effective countermeasures to its development

- Fundamental and applied research in critical infrastructure protection from terrorism

- Basic research on methods of protection of operators of critical infrastructure from "superstrong" and "superweak" external physical fields

- Basic research on methods of remote initiation of land mines

- Basic research on methods of remote detection of land mines and the electromagnetic devices managing them (radio receivers and wire lines), and methods of chemical marking of explosives

- Basic research on the vulnerability of critical infrastructure to electromagnetic terrorism

- Development of ways to protect potential terrorist targets from attack by their early detection and neutralization, using protective techniques such as the application of chemically active substances, pulses of a high pressure, or nonlethal techniques, such as optical and laser.

The USA has undertaken a coordinated program in the sphere of countering technological terrorism. In particular, in February 2003 the USA issued its first report on critical infrastructure protection, The National Strategy for the Physical Protection of Critical Infrastructures and Key Assets (Department of Homeland Security, 2003). Issued in order to comply with Homeland Security Presidential Directive-7 (Bush, 2003), this document provides clear goals and objectives to ensure the continuity of operations of key elements of American public health and safety, the economy, and public confidence. The list of major initiatives provides a road map that could be used by other nations desiring to develop similar safeguards against terrorism. Again in 2006 the Americans focused on their counterterrorism needs in the "National Infrastructure Protection Plan," issued by the Department of Homeland Security (Department of Homeland Security, 2006). It focuses on cooperation at the federal level for the protection of American "critical infrastructure and key resources" (CI/KR), inviting public–private partnerships for success. Its analysis is useful for the development of corresponding joint documents among the nations.

5.4.1. SHORT-TERM CHALLENGES FOR R&D

There are several actions that need to be undertaken in the near term. Two immediate innovative challenges are (a) recovery of cities post-CBRNE attack with a rapid cleanup capability; and (b) empowerment of the public and workforce teams to self-protect and self-rescue during sustained urban warfare.

The *cleanup challenge* for cities requires innovative cleanup methods, testing of innovative materials suitable for an urban environment, and the consideration of low cost, widely available technology and systems. One part of the solution is the application of scientific teams as rapid response support teams who move forward with the first responders. Large teams for risk assessment, survey and cleanup could be built with voluntary assistance from the private sector and supported by joint training and exercising programs.

The *challenge of protecting the public* requires considerable attention to the underassessed capability of the public and workforce teams to help themselves when under attack. These groups can be empowered in self-protection and self-rescue during sustained urban warfare. Systems for citizen self-help and work team protection are a multidisciplinary task with training programs, information systems, and special education. The goal of overwhelming terrorist capabilities can be achieved in part with preemptive allocation of small defense CBRNE capable teams within central business districts of cities and other locations, and to deny control of underground, sea or air space by preemptive assessment of vulnerabilities to critical attacks on strategic infrastructure.

Inclusion of a *Homeland Resilience Acquisitions Policy*, based on a strategic operations model, may energize the involvement of national scientific laboratories and small innovation-focused groups and deliver considerable weight to the capability of first responders. In addition, the private sector may contribute to scientific research through the commercial requirements for systems to protect the workforce and business operations. Such an acquisitions policy requires cost/effectiveness assessment and a simple method for comparison across scientific research programs (see Table 4). This enables political decisions and administrative advice to evolve, derived from a comprehensive understanding of the competition for funds. The use of scientists and equipment for surveying environments post CBRNE attack requires an acquisitions policy within government. It also requires the front line establishment of teams capable of managing a preliminary cleanup of contaminated areas or isolation of hot spots. This will mitigate the economic destruction of central business district attacks with any of the CBRNE weapons (see Table 3 for examples of CBRNE Weapons).

Critical to an effective response to all hazards, but especially terrorist attacks, is the use of an effective *command and control system*. One method that NATO members should consider is the adoption of a civilian standardized and universal command system on the European level. Common military systems allow for coordination of military assets. Similar initiatives must be undertaken if cross-border mutual aid is to be effective. In situations where there are language barriers among first responders, the use of a common command and control system will ease communication and integration of activities. But efficient and effective command and control require supporting systems, such as adequate and interoperable communications.

In Italy, which has adopted the American-developed Incident Command System, there was an event that demanded assistance from outside the country, but the various first responder agencies did not have interoperable radios, which limited the success of the response. Some of the smaller nations are further along in developing interoperable communications nationwide, but others, like Germany, are too large to afford new radios and matched equipment for all first responders. In most instances the first responders are based in local communities who bear the financial burden of providing equipment. They do not want to have to provide new radios just to match people in other parts of the nation for national response purposes, or in other nations just to prepare for the possibility of cross-border mutual aid. Technological solutions must be found to allow for the continuity of disaster response work and national mutual aid without the wholesale replacement of radios and response equipment. Systems such as wireless, broadband and new radio spectrum offer the potential for enhanced communications bridges for emergency response.

Information and Communication Technology (ICT) can assist the emergency manager to connect, to gather, to view and to work with all kinds of organizations. Each nation works on its own in the daily routine. In case of a catastrophic terrorist attack first responders and emergency managers will have to work together and have to exchange their knowledge and data. The need for Multi Agent Systems has become urgent.

Three additional challenges must be addressed through the application of R&D to communications technology:

1. Start and do something: Go on with discussing and studying the optimalization process for better preparedness, but do not use this as an excuse for delaying activities. Buy standard software and off the shelf communications bridges, and start to use them, while still conducting an in depth analysis of the real problems.

2. Connect and agree: Technology says everything is possible. So allow the telecommunications companies to show connections among each other. Wireless networks, laptop computers enabled with aircards, cellular phones, BlackBerrys and developing handheld devices like Treos (Palm, 2005) all offer the possibility of more powerful technology integration leading to enhanced information sharing and situational analysis. Interoperability is the main ICT problem case for the next years. Interoperability starts with making agreements with other organizations about the methods that will be used to create expedient interoperability, leading to fully integrated systems as budgets permit. For example, a cache of radios for exchanging at the scene of an event can start the process of interoperability for mutual aid, but technology such as WiFi and microwave-based networks can be built out to serve local jurisdictions, then regions within a nation, and finally to provide effective cross-border communications. New computer-based networks offer the opportunity for innovation leading to cheaper and more flexible systems. For a discussion of communications interoperability see Chapter 3.

3. Focus on systems development: Management has to make decisions about operational systems implementation. The shape of the command and control system will influence the ICT industry to work toward developing hardware systems to create connectivity.

5.4.2. MID-TERM CHALLENGES FOR R&D

The protection of urban space, critical infrastructure and the public from catastrophic attacks with CBRNE weapons requires R&D activities which create a *protective environment for citizens*. The opportunity for development of an overwhelming capability to minimize terrorism through a global $C4^I$ communications and operations strategy is an important goal of modeling for the middle tier timeline in Table 4. Within 10 years it should be possible for scientific applications to support global financial infrastructure protection and flexible acquisitions policies for homeland resilience. There needs to be an accelerated conversion of defense materials and technology for civilian protection.

Another second tier requirement is for a *global communications systems* for citizen, business and government preparedness, and for mitigation through combined campaigns managed through global networks established through the G8, UN and NATO. Preemptive measures that can be pulled together globally require real-time communications nets between nations and across private sector industry. The G8 nations have the possibility of providing a lead in progressive measures and a support mechanism to stabilize any region under continuous attack.

A further second tier of work is required to utilize international scientific partnerships to *overwhelm terrorist progress in recruitment and technology*. A global coordination in research and development applications can be faster and stronger in recognizing patterns of terrorist activity, and will be more effective in sharing technological countermeasures. Currently terrorist groups are operating within a global battle space and without limitations to shared data, and this may be providing a critical leverage in some conflict zones. Additionally, the Iraq War experience gained since 2003 is showing that field research and testing is moving terrorist capability well ahead of many of the defensive measures (e.g. continuous improvement of improvised explosive devices).

Networked systems for managing battle space information and operations requirements may be adaptable to soft extensions into national private sector communications. These networked systems would also be models for global systems deployment by G8, UN and NATO as terrorist alarms, for campaigns across national boundaries and for rapid response for contamination cleanup of cities. These networked systems would maximize utilization of shared technology and scientific capability between nations.

Advanced scientific measures may also be required within the 10-year period to *combat standoff delivery of weapons* from water-borne, air-borne and surface-to-surface or surface-to-air weapons, some of which may enable terrorists to deliver attacks from outside the jurisdictional control of the target nation.

5.4.3. LONG-TERM CHALLENGES FOR R&D

There are many high-tech operational solutions in defense materials and systems which can be redesigned and applied in the urban setting to assist mass casualty protection. R&D should focus on the development of new defense materials or adaptation of existing materials for mitigation of blast. Such materials are part of a proactive urban counterterrorism defense with applications in urban planning for streets, railway tunnels, and building cladding.

Within two decades, the protection of urban space, critical infrastructure and the public from catastrophic attacks with CBRNE weapons requires supportive policies for acquisition of scientific solutions to create a protective environment for citizens. This will reduce the risk for any citizen to be a victim in a terrorist attack. Most importantly, the long-term contribution of science should be toward an overwhelming capability to minimize terrorism through global C4[I] and operations strategies. The flexibility of science to work ahead of terrorist technology, strategies and tactics requires a cooperative development of programs between the national scientific laboratories and with assistance also from innovative groups of researchers. The regular assessment of risks and vulnerabilities through a Catastrophic Risk Model that is outlined in Tables 3 and 4 can provide a real-time driver for scientific goal setting and for defining pathways to applications from pure research and from adaptations of defense technology.

6. CONCLUSIONS AND RECOMMENDATIONS

6.1. International Challenges of Cooperation Against Terrorism

Modern terrorism has developed a transnational character, which then requires international cooperation in the sphere of antiterrorist activity, and this becomes an important direction for ensuring the maintenance of national safety against terrorists. Terrorism is deeply rooted in social inequality, environmentalist concerns for the exhaustion of natural resources of the Earth, and religious extremism, which are all connected to global problems of a modern civilization such as disproportionate social and economic development of the countries and regions of the world. The decision to actively manage these problems based on international norms is probably the only unifying basis for the efforts of the world community to combat terrorism.

Counteracting international terrorism demands coordination of the research efforts of scientists and experts, and cross-border coordination of actions, in order to decrease the vulnerability of developed nations to terrorism. Only robust international cooperation will lead to success in countering the terrorists' actions. Within the framework of the existing national programs it is expedient to develop an international agreement, such as "The National Plan of Reaction to Extreme Situations of Trans-boundary Character Connected to Threat of Acts of Terrorism" that was developed in Russia.

The danger that WMD may be used in future terrorist attacks is a major threat to the civilian population and society's critical infrastructure. This places special demands on the introduction of various preparedness measures. Because of its special nature and consequences, an event involving WMD will give rise to enormous media pressure and great demand for information, not only from the public but also internally within the organizations affected by the event.

Despite recent, and valuable, inputs there have been limited efforts of systematic civilian preparation against WMD. However, the world evolves, and so should domestic and international threat-mediating activities—particularly in the face of nonconventional terrorist threats. The general development of society, entailing new forms of risk and vulnerability, calls for increased alertness and continuous adjustment of security measures and preparedness, on the part not only of the administrative and political leadership, but also of various public authorities, business and industry, private organizations, right down to each individual citizen.

A difficult issue is the dual responsibilities that first responders bear toward their jobs and their families. If their families are in peril, would first responders stay on the job at the risk of their families' safety? There should

F. L. Edwards and F. Steinhäusler (eds.), NATO and Terrorism, 179–190.
© 2007 *Springer.*

be a plan to move first responder resources around to keep neutral first responders on the ground at a catastrophe. First responders of the victim community should be expected to deal with family first and then come back to work.

Individual choices will be made at the time of the event. What is the population's threshold of tolerance for an event in their community? How will they respond to direction from the government? Hurricane Katrina demonstrates that while most of the community will believe the message of credible government leaders—80% of the population of New Orleans evacuated in advance of the storm following national Hurricane Center forecasts and at the direction of Governor Kathleen Blanco and Mayor Ray Nagin—a significant proportion of the population will not or cannot conform to government requests.[1] This relatively small percentage of the population—20% in the case of the residents of New Orleans—creates a search and rescue challenge for first responders, often at the risk of their own lives, and a public relations nightmare for all levels of government. Emergency planners need to consider that first responders are accustomed to saving lives, not estimating acceptable losses.

This question also applies to the first responders. First responders are representative of their communities' populations, so what are their individual choices going to be? Will they always do what other government agencies tell them to do? The first responders in Hurricane Katrina were also victims, with 80% of the New Orleans Police Department (NOPD) becoming homeless due to either the hurricane or the levee failures. As many as 54 NOPD employees abandoned their posts to take care of their families, and were later fired.[2] Over 1,000 others stayed on the job, rescuing people who were stranded in their homes, and protecting the urban search and water rescue teams who came to help through the FEMA's Urban Search and Rescue program.[3] How can emergency planners estimate the likely role of local first responders during such a catastrophic disaster? When should mutual aid be requested? When should assistance from federal agencies be requested?

Finally, the role of the military in civil protection is an evolving question. On the one hand, the politicians tell the community that the civilian elements of civil protection are "plenty prepared," but the gap analysis suggests that this is not universally true. There is a disparity between building and maintaining the public confidence in government, and maintaining real prepared-

---

[1] For a complete discussion of Hurricane Katrina, see Appendix 5.

[2] For a detailed explanation of the impact of Hurricane Katrina on law enforcement, see Edwards (2006).

[3] For a detailed account of the role of search and rescue in Hurricane Katrina, see Chapter 4.

ness for the first responders. Civil protection is a combination of civil defense and catastrophe response, such as a response to an earthquake. The military is starting to do civil protection, as in NATO's current role in Afghanistan.

Within the EU, vanishing borders make it hard to explain to the population of each nation why national civil protection agencies are still needed. In addition, NATO's shift to civil protection creates problems with treaties. In Europe there are mixed circumstances where civil and military intersect.

The emergency management leadership realizes that the system is not perfect and needs additional capabilities. While the emergency management system is prepared and functioning, there are still many gaps that must be filled to save lives. SHAPE is improving NATO's response functions by looking at areas that need improvement. The more organizations try to improve, the more they realize what needs improving. Improvements in command structure integration and CBRNE preparedness have been carried out just in the 3 years since the NATO Advanced Research Workshop conference in Stuttgart, and 2 years since the publication of the book that grew out of it, *NATO and Terrorism: Catastrophic Terrorism and First Responders* (Steinhäusler and Edwards, 2005), but there is still more to be done to counter terrorism.

6.2. Conclusions of the Workshop

1. Terrorist groups operating in a fourth-generation warfare environment are capable of inflicting significant loss of life and damage, even in a nation that is well prepared, as demonstrated in the UK in July 2005. Since nation-states cannot engage in fourth-generation warfare, second-generation warfare techniques must be abandoned and advanced third-generation techniques adopted, such as scalable building blocks of response units, rather than monolithic and hierarchical centralized command structures.

2. Terrorist groups have the capability to acquire, or have already acquired, the means for creating and dispersing chemical, biological, and radiological weapons, as well as the means to access nuclear facilities to cause catastrophic releases.

3. EU and NATO nations currently have the capacity to respond to CBRNE events, but additional work is needed to improve the effectiveness of the response, including improved alerting and warning systems; common command and control systems and mutual aid plans; the speed and accuracy of detection and characterization of the attack; the medical and psychological care of the public; and the rapid physical and economic recovery of the community. Better personal protective equipment is

needed for first responders operating in a contaminated environment that is also very hot or very cold.

4. EU and NATO nations currently have the capacity to recover from CBRNE events, but additional work is needed to improve the effectiveness of recovery, including improved decontamination or washdown technologies for urban settings, better technologies for treating the victims of latent toxics and radioactive contamination injuries, and a more organized understanding of the sociological effects to improve the outcomes for the effected population. For example, plans should be in place for the replacement of housing in safe zones, and for the conversion to parklands of inherently unsafe areas of a community. Plans should be in place for the sealing off of areas with nuclear or heavy radioactive contamination that cannot be washed away in the short run. Better neutralizing technologies are needed for contamination of wetlands and other fragile eco-structures.

5. EU and NATO nations have law enforcement capabilities, including Interpol, to investigate, collect evidence, and build a court case against terrorist attackers. However, operating across borders poses problems for law enforcement in firearms management and importation by first responders, due to process, chain of evidence, first responder liability, and standards of case management. For examples, whose tort law will be followed? Whose penal code will control the arrest and trial, and ultimately the punishment?

6. GIS and computer modeling have the potential to enhance planning for response to terrorist attacks, allowing for the creation of maps and scenarios for likely terror attacks to be used in training, exercises, and equipment cache development.

7. Material science, engineering, and architecture have the potential to develop new materials and construction techniques that would mitigate the damage from terror attacks in an urban setting, but additional research is needed to create resilient materials and identify best practices to maximize effectiveness and cost-benefit.

8. Catastrophic disasters create a demand for search and rescue, evacuation, sheltering, and the care of special populations, including critically ill hospital patients. Current plans and technologies have proven inadequate in the face of a large-scale urban disaster, such as the hurricanes in the USA in 2005. A better understanding of the sociological effects of disaster, and the application of this understanding to emergency planning, is needed to improve response outcomes for the affected population.

9. Critical infrastructure locations, including nuclear facilities, are more vulnerable to terrorist attack than some facility owners acknowledge. Additional critical infrastructure protection is needed, such as the development of better technologies for physical protection, and for prevention or intervention against intrusion or attack.

10. Interoperability of communications and response equipment is a critical element for cross-border response to terrorist attacks. Communications are uniquely vulnerable to frequency problems and different operating standards. NATO and the EU should bring together their member states to accelerate the development of an interoperable communications system for emergency first responders that would overcome existing technological problems. Effective disaster response will require coordinated research programs and equipment acquisition programs.

6.3. Research and Improvement Program for NATO Member States

6.3.1. SECURITY THREAT ASSESSMENT

An analysis of major terror attacks in the past provides an opportunity to evaluate the response of the national emergency services and to extrapolate the requirements for emergency management to manage consequences due to such acts of *Catastrophic Terrorism*. The emergency response of national services was reviewed in three major events resembling the deployment of:

- WMK, i.e. the 9/11 attacks in New York City (2001)
- WMDi, i.e. the radiological incident in Goiania (1987)
- WMD, i.e. the two sarin attacks in Japan (1994 and 1995)

Furthermore, the results of large-scale emergency preparedness exercises conducted in Austria and the USA in 2005 were the subject of a comparative analysis.

Recommendations

1. Use a *Risk-based Emergency Response Plan*, accounting for the probability of terrorists to possess the means to carry out a specific attack, the probability for the successful implementation of the selected attack mode, the consequence of such terror attack for man and the environment, and the effectiveness of countermeasures taken during emergency response.

2. Develop workable approaches for effective large-scale evacuation in urban environments, C4^I, and operations in a radioactive environment at low temperatures and in a tropical climate

3. Increase the use of state-of-the-art technologies in informatics and loca-
    tion-aware technologies for victim tracking and care, as well as personal
    status monitors for first responders.

6.3.2. PROTECTION AGAINST TERRORISM THROUGH EMERGENCY MANAGEMENT STRUCTURES

A general appraisal of the differing dynamics and elements of intentional
and non-intentional catastrophic events showed that too often traditional
emergency response seems to be seen as a paradigm for the management of
major terrorist attacks. In reality these scenarios may differ significantly in
terms of timing, exposures, casualties, damages, media pressure, information
vacuums, and other critical features. Understanding these differences and
taking them into consideration could be vital for proper emergency planning
and responses.

At present, first responders are being tasked with coping with the issues
presented by a change to fourth-generation warfare, which thoroughly inte-
grates the civilian population with combatants, while most nation-states use
second- and third-generation command and control models. Uniforms and
rank insignia have been replaced with civilian clothes and concealed bombs.
Fourth-generation warfare consists of the opposition identifying all points of
vulnerability in a society, including media, economic forces, and terror tactics,
with combat based more on perception of events than on reality. This has
major implications for first responder training and organization in the future,
in particular for the role of law enforcement in emergency preparedness,
which has to address special challenges, including evidence collection to
meet the needs of the court systems of many nations.

Recommendations

1. Prepare for the establishment of prelocated CBRNE[2] (chemical, bio-
    logical, radiological, nuclear, extreme explosives and electronic/digital)
    capable defense teams in cities.

2. Develop rapid cleanup systems for contaminated cities.

3. Develop decentralized, scalable command and control elements that can
    function autonomously in configurations based on the event.

4. Apply global battle space management with networked solutions to over-
    whelm terrorist operations.

5. Counter psychological operations (*psyops*) as currently used for recruit-
    ment purposes by terrorists.

6. Create a sustainable system that links prevention and mitigation, defense and aftercare/reconstruction, and set up and expand a network that guaratees effective disaster management in the long term.

7. Develop a system of local self-assessments based on a target capability list, uniform resource typing, and credentialing.

8. Enhance interoperability across nations and professions through acquisition of interoperable equipment, common training, and realistic scenario-based exercises.

9. Adopt common command and control procedures and mutual aid protools to enhance cross-border disaster response support.

10. Improve the training of first responders in search and rescue operations after a major terror attack, improve their dedicated search equipment, place more emphasis on their overall discipline in an environment reflecing realistically the stress and chaos typical for the initial aftermath of such a situation.

11. Determine how liability theory can impact the fundamental professional service provision of emergency responders in the aftermath of a terrorist event. This may require certain legal or policy revisions to assist the emergency responder community so that they can more readily respond to terrorism.

6.3.3. RESEARCH AND DEVELOPMENT NEEDS IN FIELD RESPONSE TO TERRORISM

Emergency response is predicated on planning efforts put forth by state government, local communities, various agencies charged with protecting the public good, and professionals who work in response organizations. Terrorism response, especially catastrophic terrorism response, will tax the resources of government, communities, agencies, and individual responders. Planning for the potential emergency using predictive modeling is one proactive measure that can help limit, but not eliminate, the risks that catastrophic terrorism poses to all of those involved. Training, tactical awareness, infrastructure familiarity, and public support are necessary to provide a foundation for such proactive measures. In some cases the technology for actual emergency response has been given priority over the types of technology needed to effectively plan for the complex emergency response to a CBRNE attack.

In the case of a terrorist attack, the rescue of affected citizens is prioritized. However, the safety of the rescue workers must also be ensured. Therefore, it needs to be ascertained instantly whether any chemical, biological, or radioactive substances were utilized and are the cause for people's injuries, and whether secondary devices are a threat. At the same time, the early detection and identification of the risk is essential for the appropriate rescue and medical treatment of the injured.

Recommendations

1. Develop real-time measurement techniques which enable first responders to detect, identify, and monitor life-threatening contaminants in a reliable manner.

2. Develop GIS-based emergency preparation systems which can be used in a proactive planning process, imbedding preexisting data layers in order to assist in the postattack response efforts.

3. Improve predictive models, like the HPAC, which can supply what amounts to the "design basis threat" for testing emergency preparations by predicting the distribution and consequences of HAZMAT (chemical, biological, radiological, and nuclear) released into the atmosphere, and their human health effects.

4. Develop an acquisitions policy for international organizations and alliances, such as the G8, NATO, and the UN, and for national resilience programs to manage scientific research and applications in order to mitigate CBRNE[2] attacks.

5. Develop construction materials and techniques for infrastructure and the built environment to mitigate terrorism-related disasters.

6.3.4. INTERNATIONAL DISASTER RESPONSE

Current emergency preparedness concepts show two major deficits:

1. Lack of a comprehensive approach to realign security tasks, responsibilities, and capabilities with the new risk environment.

2. Continuing problems with regard to national and international interoperability.

Given the current international environment, preparing to prevent terrorist attacks from occurring, and mitigating their consequences, stands out as most important. In doing so, civilian authorities and emergency responders can borrow valuable lessons from the *concept of transformation*. This multinational, strategic, and prospective *interagency process* is aimed at continuously adapting a government's foreign and security policy instruments and decision-making processes commensurate with the needs of a dynamic environment. Since the distinction between domestic and foreign security concerns is becoming increasingly blurred, the traditional civil–military dichotomy is no longer adequate to deal with these new security risks. Adequate emergency management should ensure that emergency responders can effectively react to the new security challenges, and cooperate smoothly with other security-relevant actors. This can be achieved through coordinated efforts by science and technology (S&T), together with the industry, which can support the necessary emergency responders' transformation.

Recommendations

It is recommended to implement an International Action Plan, which consists of five actions recommended at the national level, respectively in cooperation between NATO and EU, such as:

1. Create a Joint NATO–EU Capabilities Group (NATO-EU CAP), i.e. NATO member states and EU member states are encouraged to assess the national state of emergency preparedness following a standardized protocol.

2. Create a Coordinated NATO–EU Training Program (NATO-EU PRO), i.e. NATO member states and EU member states are encouraged to standardize education and training of emergency responders.

3. Create a Joint NATO–EU Homeland Security Clearing House (NATO-EU SEC), i.e. NATO member states and EU member states are encouraged to jointly analyze lessons learned from major homeland security operations.

4. Create a Coordinated NATO-EU Research Program (NATO-EU RES), i.e. NATO member states and EU member states are encouraged to coordinate research related to the needs of joint capabilities planning.

5. Create an Office for Security Science and Technology (OSST), i.e. a national institution, integrating all public and private stakeholders dealing with R&D in homeland security.

6.4. Managing a Radiological Terror Attack: A Paradigm for Transformational Emergency Management

Emergency management after a radiological terror attack has to account for the large number of variables characterizing the geographical location, climatic conditions at the time of the attack, contamination vectors, and the social classes involved—all of which determine the resulting impact on humans and the environment. However, in general, three phases can be differentiated: initial response, intermediate response, and long-term response.

The *initial response phase* suffers from insufficient information: lack of knowledge of the size of the area affected, number of persons contaminated and/or injured, and the type of contaminating radioactive material. This will require a unified command (federal, state, city) from the onset, staying in the area for the complete intervention process. Within the unified command a lead agency should (a) prioritize lifesaving actions; (b) categorize the victims and sites suspected to be contaminated; (c) prioritize decontamination tasks and techniques; (d) designate an official communicator.

The provisional readiness of medical facilities and the qualification of medical staff will determine the fate of the victims. Therefore, triage has to be undertaken quickly, sorting out the relatively small number of people having received a significant dose.

The *intermediate response phase* will have to address the implementation of exceptional procedures, such as:

- Acquisition of materials and equipment needed for treatment of victims, survey, and search operations
- Hiring of expert services unavailable in the area
- Disposing of contaminated materials
- Granting legislative power to isolate contaminated areas
- Enforcing evacuation rules
- Retrieving and confiscating contaminated goods
- Demolishing contaminated buildings
- Selecting areas to store contaminated waste and debris temporarily
- Establishing sites and coordinating relief for the victims

The *long-term recovery phase* is characterized by the establishment of socially acceptable and practically applicable clearance levels for the large amounts of material resulting from the cleanup operations with low levels of residual radioactivity, i.e. clearance of buildings for any purpose (reuse or demolition), clearance for demolition only, and clearance for building rubble.

Since many of these issues have not been addressed sufficiently among NATO members, it is proposed to consider the creation of an international "Center of Excellence," perhaps under the auspices of NATO, where short- and long-term effects of radiological terrorism on society can be the subject of a coordinated R&D effort. Topics to be addressed range from the effects of radiological terrorism on business, real estate, and agriculture to behavior of the public (e.g. refusal to return to decontaminated areas, radiation phobia). Such a center could also develop models and algorithms, for example, cost models for cleanup operations (procurement of general equipment and material, dismantling of contaminated sites, waste treatment and disposal, security, surveillance and maintenance, site cleanup and landscaping, project management, engineering and site support, fuel, other costs).

6.5. Conclusions and Next Steps

NATO and the EU have an opportunity to create a partnership that can develop plans for the ultimate defeat of transnational terrorism. Professor Amos Guiora, a retired Israeli judge advocate general, has warned that the opposition to terrorism by the world's liberal democracies is not a "war" but a state of being for the foreseeable future. Heiko Werner has warned that nations cannot afford to maintain the civil defense stance of an "eastward-facing war-ready society." Dan Goodrich has warned that fourth-generation warfare is being practiced by the terrorists, and that while liberal democracies cannot tolerate their militaries undertaking this "no holds barred" approach to warfare, they must embrace third-generation warfare if they have any chance to succeed.

The Science, Technology and Society Conference on National Armaments Directors (STS-CNAD) Workshop of 2006 brought together experts to explore the needs of first responders in a terrorist attack, and to anticipate the work that needs to be done in the sciences and social sciences to obtain the greatest benefits for society from current knowledge. They also assessed the needs of the first responders for better structures and technologies for the protection of lives, the environment, property, and civil society.

Recovering from the horrific probable CBRNE attacks requires another serious evaluation. How can nations prepare for the long-term recovery and rehabilitation of their societies after a truly catastrophic event? Hurricane Katrina is a model of what a natural disaster can do when a large area—90,000 square miles—is impacted in a short time. But hurricanes go away and flooding stops eventually. Radioactive materials have a half-life of centuries. Biological organisms let loose on a naive population can circle the globe for decades, wrecking havoc as populations succumb to the disease, and die or become permanently disabled. Chemical events can leave facilities and sections of communities uninhabitable for generations due to stigma and fear.

NATO should undertake thorough investigation, study, and policy development for the long-term responders and their roles and needs. Public works, urban planning, financial and economic development, housing, public health and mental health professionals should gather to intentionally examine the true impacts of a catastrophe on a modern urbanized society, and develop strategies for R&D, applying lessons learned, and creating a new resiliency in the members of the alliance.

Finally, NATO should study how the alliance can reframe its international relations to undercut the recruitment to, and the financial support for, terrorism. Foreign aid, peacekeeping, and economic development opportunities may be the ultimate weapon against terror, and against the potential destruction of liberal democracy in the world.

APPENDIX 1: LISTS OF PARTICIPANTS

NATO Joint STS-CNAD Workshop
Emergency Management after a Major Terror Attack:
The New Challenges for First Responders and Civil Protection
Ericeira, Portugal, March 2nd – 4th, 2006

Name	Position	Affiliation	Address	Country	Phone/fax/e-mail
Baciu, Adrian	Coordinator	Bioterrorism Unit, Public Safety and Terrorism Sub-directorate, Interpol, General Secretariat	200 quai Charles de Gaulle, 69 006 Lyon	France	a.baciu@interpol.int
Ballard, James David	Associate Professor	Department of Sociology, California State University, Northridge	18111 Nordhoff St., Northridge, CA 91330-8318	USA	818-677-2009, 818-677-2059 ballard@csun.edu
Benigno, Alexandre		National Service for Fire and Civil Protection	Av. Loureiro 192, 2775-599 Corcovelos	Portugal	abenigno@snbpc.pt
Borko, Sasa	Policy Officer	Civil Protection Unit EU Commission	Brussels	Belgium	+32-2-298-1061 sasa.borko@cec.eu.int

191

F. L. Edwards and F. Steinhäusler (eds.), NATO and Terrorism, 191–194.
© 2007 Springer.

Name	Position	Affiliation	Address	Country	Phone/fax/e-mail
Bretschneider, Günter	Head	Euro-Atlantic Disaster Response Coordination Center (EADRCC), NATO	Boulevard Leopold III, B-1110 Brussels	Belgium	+32-2-707-2670; +32-2-707-2677 EADRCC@hq.nato.int
Butterfield, Alan		Supreme Headquarters Allied Powers Europe (SHAPE), NATO	Brussels	Belgium	alan.butterfield@shape.nato.int
Campeao, Alvaro	Head of Emergency Planning Division	National Service for Fire and Civil Protection	Rua dos Pedrogos n° 11 – Pedrogos, 2640-567 Mafra	Portugal	+351-91959419 ACampeao@snbpc.pt amkcampeao@sapo.pt
Edwards, Frances L.	Associate Professor	San José State University	20405 Via Volante, Cupertino, CA 95014	USA	kc6rhm@yahoo.com
Goodrich, Daniel C.	Research Associate	Norman Y Mineta Transportation Institute	20405 Via Volante, Cupertino, CA 95014	USA	+1-408-807-0930 Rule308ocs@yahoo.com
Leivesley, Sally	Managing Director	NewRisk Limited	Frazer House, 32/38 Leman St, London E1 8EW	UK	+44 2086603873; +44 2087638603 Sally@newrisk.com
Maerli, Morten Bremer	Senior Research Fellow	Norwegian Institute of International Affairs	PO Box 8159 Dep., N-0033 Oslo	Norway	+47 2299 4107; +47 900 74420 (cell.); +47 2236 2182 MBM@nupi.no

Name	Position	Affiliation	Address	Country	Phone/fax/e-mail
Mouraux, Sophie		SGDN Protection et Sécurité de l'Etat	51 blvd de la TOUR, Maubourg 75007, Paris	France	01 71 76 85 87 Sophie.mouraux@sgdn.pm.gouv.fr
Mullendore, Kristine B.	Associate Professor	School of Criminal Justice, Grand Valley State University	271 C DeVos Center, 401 W. Fulton St., Grand Rapids, MI 49504-6431	USA	+1-616-331-7147; +1-616-331-7155 mullendk@gvsu.edu
Otten, Jan	Director	Respond BV	Jules Verneweg 121, 5015 BK Tilburg	The Netherlands	+31.13.5321001; +31.62.605.2839 Jotten@respond.nl
Potapov, Boris	Leading Scientist	EMERCOM of Russia	Moscow	Russia	007-8-916-389-13-55 potapovb@mail.ru
Rechenbach, Peer	Deputy Chief	State Fire Department and Emergency Medical Service of Hamburg	Westphalensweg 1, D-20099 Hamburg	Germany	+4940-42851-4002; +4940-42851-4009 peer.rechenbach@feuerwehr.hamburg.de
Rydell, Stan	Radiation Scientist	US Environmental Protection Agency	1 Congress St., Boston, MA 02114-2023	USA	+97205778407l; +97289725056 sym42@walla.com
Sagiv, Yossi	Jerusalem District Commander	Home Front Command, IDF	MOSH Dyen 36/2 Modlin, Jerusalem	Israel	+97205778407l; +97289725056 sym42@walla.com
Schapelhouman, Harold	Special Operations Division Chief	Menlo Park Fire Protection District, California Task Force 3 – National Urban Search and Rescue Program	300 Middlefield Rd, Menlo Park, CA 94025	USA	+1-650-688-8426 harolds@menlofire.org

Name	Position	Affiliation	Address	Country	Phone/fax/e-mail
Steinhaeusler, Friedrich	Senior Scientist	Oppenheimer Scientific Inc.	101 Continental Blvd., Suite 1660, El Segundo, CA 90245	USA	+43-662-8044-5701; +43-662-8044-150 Friedrich.Steinhaeusler@sbg.ac.at
Trapp, Ralf	Senior Planning Officer	Organization for Prohibition of Chemical Weapons (OPCW)	2517 Jr. Joh de Wittlaan 32, The Hague	The Netherlands	+3 70 4163770 Ralf.Trapp@opcw.org; Ralf.Trapp@gmail.com
Urbanek, Jiri F.	Associate Professor	Faculty of Economy and Management, Department of Civil Protection, Czech University of Defence	Kounicova 65, 612 00 Brno	Czech Republic	+420973443913; +420603326355 (mob.) Jiri.Urbanek@unob.cz
Werner, Heiko	Head	International Project Division, German Federal Agency for Technical Relief (THW)	Deutschherrenstr. 93-95, D 53177 Bonn	Germany	+49 173 295 1885; +49 228 940 1144 heiko.werner@thw.de
Zaitseva, Lyudmila	Affiliate Researcher	NATO Joint STS-CNAD Workshop Secretariat, Division of Physics and Biophysics, University of Salzburg	Hellbrunnerstr. 34, A-5020 Salzburg	Austria	+43-662-8044-5701; +43-662-8044-150 Lyudmila.Zaitseva@sbg.ac.at

APPENDIX 2: COUNTERING FOURTH-GENERATION WARFARE WITH FIRST-THROUGH THIRD-GENERATION WARFARE: A THEORY OF FUNCTIONAL FORCE EMPLOYMENT FOR FIRST RESPONDERS[1]

Background

Over the past few years there has been an increased collaboration between military assets and intelligence and civilian emergency responders in the field concerning planning and training for situations involving WMD. Since the 1997 creation of the Metropolitan Medical Strike Teams (Nunn-Lugar-Domenici, 1996), training for civilians has often been modeled on military doctrine, tactics, and equipment. Often overlooked are the significant differences between the military's mission and that of civilian first responders.[2] A clear understanding of what is being considered for adaptation from the military by civilian organizations, be it tactics or equipment, is critical to the development of appropriate civilian field-level response plans. For example, fundamental considerations for the military's operational or equipment selection are based on policies that permit, even demand, "acceptable losses" of service members' lives in certain situations. Civilian authority, in contrast, could never consider planned loss of life as acceptable under any circumstances. That divergence of mission and policy needs to be kept in the forefront of any comparison between civilian and military operations,[3] including the study of how "generational warfare" impacts the organization of civilian response to terrorism (Table A-1).

Two Dominant Military Theories

In current military thought two principal theories exist regarding what wins wars. The first is "mass and materiel" and the second is "force employment." With mass and materiel the guiding principle is to have newer technology

---

[1]Material for this Annex is based on a paper presentation by Daniel C. Goodrich delivered at the NATO STS-CNAD Workshop, Portugal, 2006.

[2]For the purposes of this paper, "first responders" means police, fire, emergency medical services, public works personnel, and other civilians who go to the scene of a disaster for the purpose of rescuing and treating the victims and mitigating the attack.

[3]For a complete review of the military and civilian response, see Larsen and Peters (2001).

F. L. Edwards and F. Steinhäusler (eds.), NATO and Terrorism, 195–220.
© 2007 *Springer.*

and information-gathering capability than the enemy, and use a preponderance of accurate firepower to eliminate potential threats before they are in a position to return fire. The individual's contribution is a part of a larger system that enables their replacement, or loss, with little or no impact upon the capability of the unit to which they are assigned. Equipment, not personnel, is considered principal in assuring victory. Advancements in technology should therefore be embraced, and modifications to operational doctrines made to maximize their effectiveness.

The second theory, force employment (Biddle, 2004), takes the opposite position. People, and how they are utilized, are considered to be the principal reason for victories. Equipment is considered to be a force multiplier of the basic capabilities of the personnel, instead of the other way around. The basic point is that war is chaotic and highly susceptible to change. Equipment failures, new weapons systems and tactics employed by the enemy, and other factors contributing to friction must be overcome, no matter how detailed the plan. Therefore, the human ability to adapt and overcome is a more important consideration in ensuring eventual success. This does not imply that a minimal amount of force would be used to overcome the enemy. It may actually utilize more than the mass and materiel theory, but for a different purpose. Use of redundancy as a contingency is one such option. The force employment approach, however, mainly attempts to utilize every resource at its disposal, even if in an ad hoc fashion, in achieving its goals.

In reality both of these theories are used in actual warfare and share a symbiotic relationship. Advancements in technology do not take place in a vacuum. They occur due, in part, to adaptation and improvisation of individuals contributing collectively. Likewise, the complexity of modern weapon systems does demand subservience of individuals in order for them to be coordinated and function properly. While civil authority shares, to a certain extent, the same situation, there is a clear planning focus on force employment in first responder agencies.

TABLE A-1. Generations of warfare

Historical references	Generation of warfare			
	First generation	Second generation	Third generation	Fourth generation
Timeframe of dominance (military)	1648–1861	1861–1918	1918–Current	1924–Current
Battlefield examples	Napoleonic campaigns; line and column formations	American Civil War; World War I; trench warfare	World War II; "Blitzkrieg"; 1967 Six Days War; combined operations; manoeuvre	Avoids conventional concept of "battlefield" based on geography
Tempo of operations	Set paced operations required to coordinate various assets for offense as well as defense	Set paced operations for offense; layered defense used to disrupt attack and "buy time" to develop counterattack	Flexibility and adaptability stressed to exploit opportunities presented in both offensive and defensive operations	Deliberate avoidance of opponents' tempo of operations range by alternatively slowing down and speeding up
Transportation	Foot; horse; sail	Railroad; steamship	Car; aircraft; ship turbine	Exploits all available resources
Social	Strong upper- and lower-class representation; middle class represented only in guild/trades	Developing broad-based middle class	Strong middle-class representation	Religious; tribal; clan

Historical references	Generation of warfare			
	First generation	Second generation	Third generation	Fourth generation
Political	Monarchy	Democracy; constitutional monarchy	Liberal democracies; nationalist	Religious; patriarchal/matriarchal; tribal; clan
Educational	Low literacy rate	Mid-level literacy rate	High literacy rate with technical specialization	Variable
Economic	Agrarian; light industrial	Full industrialization	Post industrialization	Variable
Communications	Flags; signal mirrors; runners	Telegraph; telephone	Radio; digital	Variable
Command	Narrow; few junior ranks present	Pyramid; junior ranks present but used for specific tasks	Flat with strong development of flexibility in junior ranks	Independent
Control	Centralized	Centralized; based upon skill or mission	Decentralized; based upon objective	Independent
Directives	Task-specific; little or no room for interpretation	Plan-specific; variance from identified task permitted if situation dictates	Goal-oriented; plan is flexible and based on resources available and situation	Autonomous
Mission training	Rudimentary skills only; primary knowledge retained by command	Some primary knowledge imparted to mid-level personnel but focus remains on skills	Detail-oriented; primary knowledge imparted to junior personnel	Variable

Current characteristics	Generation of warfare			
	First generation	Second generation	Third generation	Fourth generation
Personnel turnover	High; requiring frequent training in skill development	Average; making skill development an ongoing issue but some primary knowledge imparted	Low; permitting concentration on primary knowledge once skills are mastered	Clan; family; tribal; very low or no turnover
Complexity of mission	Rarely deviates from a known standard and has a well-developed operational procedure	Can deviate from a known standard; some ability to adapt but limited	Extreme deviation likely requiring frequent adaptation	Dependent upon capability; loose alliances may be formed with others if goals are similar
Supervision	Constantly required to insure scope of work is being addressed	Some supervision required but only on more complex tasks	Required only when objective needs clarification or situation makes objective impossible to obtain	Little or no supervision

TAILORING TO CIVIL AUTHORITY

Civil first responders are vested with a broad mandate to protect the public. Their resources are limited or generic, necessitating frequent improvisations and adaptations. Training is therefore considered paramount, with equipment a secondary factor. In that respect, the force employment theory is a better representation of their circumstances, and should be noted as the underlying theory of this paper. Personnel, not equipment or procedures, are the most important asset for civilian organizations. Equipment and procedures therefore augment personnel to increase their capabilities (force multiplication). The focus is how to maximize the human potential that forms the basis of the effort. In order to accomplish this task among varieties of organizations, a common framework of comparison is required. Of all the theories that complement force employment, the Generational Warfare (GW) model (Lind *et al.*, 1989) provides the necessary latitude for adaptation to civil use as a theoretical basis. It also assists in understanding the core command issues of joint operations by concentrating on functional capability.

STRUCTURE VERSUS FUNCTION

Organizational structures, mainly hierarchical, have existed for millennia. Span of control, specialization of assets, and chain of command are used in militaries, civil organizations, and private enterprise. Each organization has key points of reference that theoretically should permit their merger with other similar entities. If that were, in fact, the case, there would certainly be no need for exchange programs among various military institutions, both inside a country and with its allies. Because transparent mergers are not possible, exchange programs exist to enlighten the participants about how an organization *functions*. Organizational structures provide the necessary framework of reference points, allowing an outsider to use the proper protocols and navigate through the organization. When placed under stress, a viable entity will rely progressively less on organizational structure in favor of what is actually *functional*. Often these are called "reorganization" or "transitional" phases that the organization will cycle through. A prime example of this is the transition from a peacetime military to a wartime military in which the country, or countries, involved reorganize to address the inherent differences between the two environments, though the organizational structure may remain the same.

There can be a distinct difference between how an organization is structured and how it actually works. The reason can be found in how it functions based on social factors and mission requirements, as well as transitions over time. Imagine placing a Navy attack submarine commander in an Air Force tactical fighter group. It can be, and probably has been, done, but with the commander as an observer. Think, though, of placing that commander in a command position within the fighter group, based solely on military rank. Command problems stemming from differences between fighter and submarine strategic and tactical thought become evident. For example, in a submarine everyone knows everything about the hull and the safety procedures, and in an emergency each crew member can take action to stop the emergency without further orders. On the other hand, a tactical fighter group is constantly reevaluating its actions based on its current mission and the obstacles that it confronts. These different approaches to work create incompatibilities in terminology, operations, and organizational structure. Even if the commander is kept strictly in a managerial role, existing conflicts in terminology and training would result in functional impacts upon the organization, due to how the commander thinks or processes information, particularly under stressful conditions, and how the tactical fighter group would be operating.

In the civilian arena, a similar situation exists between most law enforcement and fire departments. It is how an organization functions that results in the selection of certain individual attributes to emphasize in recruiting new members, and those in turn contribute to a collective consciousness of how the organization perceives and expresses itself. One comment from a local firefighter states the issue quite eloquently: "The job picks you, you don't pick it." As such, law enforcement and fire, two organizations that have similar organizational structures (and may even have identical mission requirements in some cases) working together, may result in loggerheads over functional differences. This is due not only to individual personalities, but also to the climate that fostered and encouraged the attributes of those personalities. For example, law enforcement encourages individualistic action, while fire requires multimember teams. It is through function that real operational capability and compatibility exist, and where "Generational Warfare" theory offers a theoretical framework for analysis.

Generational Warfare

The theory of "Generational Warfare" (GW) is grounded in sociological, economic, political, and technical evolutions that result in fundamental changes in how a country organizes and prosecutes its conflicts. The basis is the premise of the nation-state, indicating a clear relationship among the government, military, and population, recognizing their dependence upon one another. Therefore, the underlying values of all three share common grounding that complements each other. The lessons garnered from the military, in particular how it works functionally, can thus be observed for comparison in civil and other organizational structures.

The first three GW models listed below are still in use in the world, and are a reflection of the parent country's social, economic, and geographical conditions, as well as the specific agency or department's mission. The terms are not meant to insinuate inferiority on the part of any nation or organization. The theory is intended to provide a context for comparisons between the functional mechanisms of a country's military and its civil first responders, as well as to other countries' military and civil first responders. As will be noted in the GW descriptions, there are other factors determining why an organization might be employing the tactics and systems of a specific "generation" of warfare. The focus should remain on the terminal purpose, that of enabling a mechanism that will allow different responding resources from various countries to function in a coordinated fashion. This will minimize overlap and gaps to achieve a successful resolution to a catastrophic event, even while utilizing equipment and techniques that may be less than optimal. It should also be noted that none of these generations sprang forward fully developed. In some cases, the process evolved over centuries, in others mere decades. Current organizations can also be at different developmental levels within a particular generation. This is more evident in the newer generations than in the older. These descriptions are therefore intended to convey the process of further development also.

FIRST-GENERATION WARFARE

First-generation warfare (1GW) is identified as being established with the Peace of Westphalia in 1648. It dominated the period between 1648 and 1861. While most perceive this period as the time gunpowder came to control the battlefield, a broad perspective shows that much more was occurring. The religious changes, availability of printed books, greater literacy, rise in nationalism, and an increasing diversification and specialization of the workforce all worked to push the ability to wage war almost exclusively into the realm of the governments that managed their respective nations. The very chemistry, metallurgy, engineering, and study of ballistics to place guns on the battlefield required multidisciplinary access that private, and even religious, organizations could not afford, or philosophically support. Military organizational structures became progressively more complex as increasingly larger armies were fielded. Although they may have been largely conscripted, the personnel comprising the ranks were citizens of the nation their army represented. In order to ensure rapid communication of orders and instill instant obedience, and to use moral force as a coherent element to hold the unit together, the units were deployed in a strict order known as "line-and-column formations." It would take another century and a half before this crystallized with Napoleon's recognition that the population has a direct connection to the military, forming a triangle among the government, military, and population in the wake of the French revolution (Van Creveld, 1991). Command and control, though, was still relegated to a few professionals, and relied principally on line-of-sight communications (flags, mirrors, and runners) to coordinate resources. This placed a premium upon set-paced battle plans, where rapid adaptation to changes on the battlefield was not possible. If an opportunity for exploitation of an enemy's vulnerability presented itself to a particular unit and was pursued, the surrounding units relying on its support as outlined in the battle plan could be put at risk. Likewise, a unit that failed to achieve its battle plan objective could induce vulnerability on their own side.

By the beginning of the American Civil War these forces became harder to coordinate and support under their narrow command and large logistical systems. The increasing lethality of the weapons used necessitated a change as well. The line-and-column formations needed to be dispersed in order to increase probability of survival, because the newer weapons were more accurate and had a greater rate of fire. Command and control needed to decentralize and new tactical approaches were required as the physical area of the battlefield increased.

1GW Relevance

The reason that 1GW methodology persists to this day is a combination of high turnover of personnel, few professionals, few opportunities for training due to cost, a mission that is extremely simple, and/or straightforward use of the technology present. The organizational approach is kept simple, led by a few professionals who have the technical knowledge, and augmented by many who have little or no experience. As such the professionals must maintain direct control of their subordinates or risk functional fragmentation of their unit. Demands upon communications technology are kept low, because few personnel have the mission-essential knowledge necessary to exploit them as assets. Therefore, few actually require communications assets in order to coordinate the resources present.

The close-order drill portion from line-and-column formations has been carried forward into present indoctrination training of not only military institutions but also civil first responders. This is due to the initial need to establish positive control over subordinates, and as a reinforcement tool for representation of the organizational command structure.

SECOND-GENERATION WARFARE

Second-generation warfare (2GW) began in the American Civil War. The period of its dominance was 1861–1918. Use of technology became a more important element of warfare. Rail and telegraph enabled the rapid movement and coordination of resources over broad areas. Weapon systems increased in range, accuracy, and rate of fire. Dispersal of personnel over a wider area and use of earthen works enabled survival, but more importantly, a greater understanding and appreciation of individual roles developed. This helped compensate for the loss of moral force or cohesion and rigid command that line-and-column formations imposed in 1GW. Command was thus flattened to a certain degree out of necessity, enabling tactical commanders to exercise some tactical discretion. What truly enabled this, however, was the increase in the number of persons with higher educational levels. Social factors, including the development of a middle class associated with the industrial revolution, were as much a factor on the battlefield as the constant improvement in weapons technology.

While command may have flattened somewhat, control did not. Requests for authorization or support from subunits demanded going up, across, and down the appropriate command structures. Although a drastic improvement over the 1GW model of fixed battle plans, 2GW methods resulted in actual time delays while awaiting a response from higher levels of command, resulting in the inability to rapidly exploit opportunities presented by the enemy's tactics and failures. It did, however, work well in fixed defensive situations. By 1917, 2GW used a layered defense to slow an enemy attack. While artillery was concentrated on the breach, reinforcements were brought in via rail from other sectors to shore up the existing defensive line or establish a new one. The stratification of different unit types into separate command elements resulted in inflexibility in the offense, forcing the fixed battle plan approach of 1GW to enable coordination. The result was attrition with reliance on massive firepower in an attempt to induce the other side to surrender.

2GW Relevance

This system is the current norm when you consider the concentration of specific resources in current military institutions. The reason is that centralized facilities reduce cost while maximizing training and use of task-specific command structures to ensure a uniform and systematic employment. The downside is that this results in little cross exposure between tactical assets, which could enable identification of gaps between various elements and result in plans or tactics being developed accordingly.

The same holds true in the civil sector as law enforcement has its own training and operational procedures as does fire. Routine assignments leave little room for interaction with other disciplines, as areas of responsibility are clearly defined. This becomes an issue when a situation presents that demands coordination between disciplines, and friction develops because of unfamiliarity. Communications assets are relied upon greatly and can be quite sophisticated, but generally compartmentalize the various organizations deliberately in order to maintain functionality within the existing command structure under normal circumstances. A partial solution to this issue is the use of liaisons at various mid-level positions to function as go-betweens or brokers for the various disciplines, in order to coordinate resources. This approach presents other problems, as the liaisons may find themselves in positions in which they can understand the issue, but lack the ability to articulate the circumstances to their command in a meaningful way. They may also have their requests overridden by their own organization due to their discipline's perception of the situation.

THIRD-GENERATION WARFARE

Third-generation warfare (3GW) began, remaining with the original classifications, in 1918 (Gudmundsson, 1989). Some would argue that it began with the rise of the German General Staff, particularly with Field Marshal Moltke. Others would argue that it was the Spanish Civil War in 1936. The year 1918 is identified because battlefield conditions of World War I necessitated a change in the approach used in offensive warfare between nation-states. In order for that to occur certain functional issues needed to be addressed. Some elements had been present well before in personal leadership styles, but other elements did not present themselves until the military realized that a fundamental operational and organizational change was required.

Higher levels of education and national identity, as well as technological developments, also needed to occur to create a new "generation." 3GW is no exception to this rule. Further complicating this matter is that it was not until the mid-1950s that a significant element of 3GW was identified. All four "generations" of warfare are subject to this debate over when the evolution to them was complete. There is invariably a time span of decades as various factors in politics, technology, and sociology need to come into play, leading to the emergence of a new warfare generation.

By 1917 it had become apparent to both sides that the existing command and control elements in offensive warfare were functionally insufficient. As a result, experimentation took place. What ensued was the creation of what became known as infiltration, or Stormtroop, tactics. While most believe

Germany held exclusive domain over these tactics, the Canadian engagement at Vimy Ridge is also an example. The key of these tactics was not only a flattening of the command structure, but also a clear understanding among all personnel as to what the ultimate objective of the operation was. Additionally the establishment of direct lines of control, particularly for artillery, among the various elements involved broke with the 2GW model. For the Canadians this entailed not only full-scale mock-ups and weeks of training and rehearsals that resulted in everyone knowing their individual assignment, but also their organization's responsibility, therefore allowing adaptation. The Germans, in contrast, relied on the ingenuity of their tactical commanders, and on the specialized training of a few personnel, known as Stormtroops, to utilize their available resources and develop their own plan and contingencies as they saw fit. Only the main focus of effort or center of gravity, *schwerpunkt* as the Germans refer to it, was identified by command. This combination allowed units, all the way down to the individual combatant, to readjust their course of action autonomously in support of that objective.

What enabled that transition goes far deeper than the battlefield. Greater individual educational levels, a breaking down of social caste systems, and an increased sense of nationalism all contributed. The increased individual firepower brought about by mass production and adoption of different weapons systems (flamethrowers, trench mortars, light machine guns, and light artillery) resulted in a highly flexible arms capability, able to rapidly exploit opportunities that presented themselves, as well as provide an elastic defense that required fewer resources than the 2GW model to be effective.

At the end of the war, Canada reverted to 2GW due to training and cost constraints. Germany, however, was able to advance 3GW even further because of post-war treaty-based restrictions on the size of their military that forced them to place a premium on quality. The introduction and use of an operational level command system to coordinate the resources and units in a given area and provide identification of the *schwerpunkt* became the norm. Detailed explanations and mission assignments were replaced with an intent style mission order, or *auftragstaktik,* (also known as directive control). This freed the tactical level to plan and implement its resources as it saw fit, in direct action or supporting operations in relation to the primary objective. This again placed a high demand upon the individual's intelligence, education, and resourcefulness. Incorporation of reliable armor and air assets into this system enabled a combined arms capability that permitted rapid adaptation to changing circumstances on the battlefield. As a result, during World War II the German military was able to achieve a disproportionate number of victories in both offense and defense despite routinely being materially outmatched.

Time as a Weapon

The critical point missed by the Germans was the principle of time as a weapon on the battlefield. Consider for a moment that the German invasion of France in 1940 went so well they thought it was a trap. The subsequent halt of offensive operations allowed the successful evacuation of the British at Dunkirk. What had developed was a command system that naturally exploited time. Without recognizing its full potential, its full power was not realized. This fact had been missed by most of the participants using 1GW and 2GW, and had contributed to their succumbing to later GW models. It was not just the equipment, technological evolutions, or tactics that delivered victory; it was an organization functionally able to control the tempo of the battlefield.

Time is a tactical advantage. What this implies is being able to move at a rate that puts your opponent at considerable disadvantage. In some cases it is slower than the enemy is capable of operating, as in an ancient siege of a city lasting years, or slow-flying low-altitude aircraft like the Swordfish pitted against the high-speed antiaircraft tracking systems of the battleship Bismarck. In most situations, though, it is the ability to transition quicker than the enemy that gives the advantage. Imagine for a moment a ballet in which the music is played too slow or too fast. The dancers will not be able to compensate, and the ballet will tear itself apart. It is the same on the battlefield for the side that does not have control over the tempo. It was not until a comparison of performance between the MiG-15 and F-86 jet fighters during the 1950–1953 Korean conflict was made that a solid theory emerged as to just how critical this issue is.

OODA Loops

Superficially, a MiG-15 is more than a match for the F-86. (Lind, 1985) It is smaller, more nimble, has a higher thrust to weight, and carries cannon. However, it lost far more frequently in dogfights to the F-86. The cause for this anomaly was that the F-86 physically placed the pilot in a position to better observe his surroundings, and therefore more consistently orient himself in relation to his opponent in the MiG-15. By making a series of moves, each putting the MiG-15 at greater disadvantage, the F-86 would finally win. An entire theory emerged and contributed to the final piece of the 3GW equation. John Boyd, a US Air Force pilot, conducted the study, and the resulting process of engagement became known as OODA loops or "Boyd cycles."

OODA stands for Observation, Orientation, Decision, and Action.[1] The reference to it being a loop, or even a cycle, is misleading. (Hammond, 2001) The theory is that everything an individual hears, sees, tastes, feels, and smells is observed. The Observations are then filtered in the Orientation phase based on previous experiences, religious beliefs, education, social background, preferences, prejudices, virtually everything that makes us psychologically who we are. This explains why you can have two individuals observe the same thing, but have radically different perceptions of what they observed. While the orientation is occurring, however, the participant is part of the observation, so an automatic loop between the two commences and continues. Once enough information is distilled by Orientation, choices for Decision are presented. For those with a great deal of experience, and thus a robust Orientation capability, the list might be long; for others, it will be short. Even here, though, sub-loops emerge back to Orientation and Observation because of constant comparisons and being part of an ongoing process as it unfolds. From the list of options in Decision, an Action is chosen and implemented. The process then loops back to Observation and continues as the repercussion of that Action, then influences Observation, and the process repeats again and again and again.

The reason for the relevance of this process is that it is not only applicable to individuals, but to organizations as well. When applied to warfare, it is referred to as "getting inside" the opposition's decision cycle, thereby placing them in a situation of increasing irrelevance and confusion, ultimately resulting in their collapse or failure. The actual use of this combination came to devastating effect in the 1967 Six Days War. Consider for a moment the imbalance of material resources, both quantitatively and qualitatively, against Israel. Some will cite that the Israelis had either better intelligence gathering or superior training to make up the difference. Neither of these, however, can explain why Israel won as decisively as it did, without addressing how tempo was exploited. From the moment Israel commenced operations until they were halted, the Egyptians and Syrians were placed into an operational tempo they could not keep up with or compensate for. The result was devastating and clearly lopsided in favor of the Israelis. Another example of dominating the tempo would be the 1991 Gulf War, but the use of superior technology against the enemy by the coalition forces makes a comparison more difficult.

---

[1]OODA Loop diagram is available at the following URL: http://www.belisarius.com/ modern_business_strategy/shay/ooda_loop_sketch.htm

3GW Relevance

Third-generation warfare is the most complex, and therefore the most costly model employed, not from a material perspective, but from the cost of training. A high educational level, a standard social background enabling common frames of reference, and a strong sense of organizational identity are needed to support such a system. Only Israel, Germany, and the US Marine Corps are known by the author to employ this approach operationally. In the case of both the Germans and Israelis, it is not limited to their militaries; their civil response systems also utilize this functional approach.

For Israel this extension is logical due to its strong military–population–government bonds resulting from its geographical position among unfriendly neighboring countries. Israel has often demonstrated an ability to adapt militarily, as well as socially, to changing threat circumstances, and is routinely used as a model of response to terrorist acts within the country.

Germany's civil authority experience is rooted in wartime lessons in firefighting and Urban Search and Rescue (USAR) operations. The adaptation from their military was a natural progression that proved functional enough to maintain.

The Marine Corps has always had a strong sense of organizational identity, requiring a high school education for enlistment, and using Boot Camp as a social structuring tool to establish their own frame of references. The inherent effectiveness of such a system rests primarily in the sociological sphere, making it problematic to employ in organizations with high turnover rates or few training opportunities. Surprisingly, though, communications assets in such a system are actually utilized less than in 2GW, as everyone automatically coordinates off one another's actions in relation to the situation, without needing to go through their respective command structures.

What makes this system particularly unique is the ingrained security it also affords its practitioners. The fluidity of action masks the command structure, and results in difficulty for the untrained observer in understanding what is actually occurring. To this day, most do not understand the German or Israeli approach to warfare. In contrast, 1GW and 2GW models are far simpler to analyze and comprehend. Another benefit of 3GW is the inherent redundancy and ability to compensate for loss of major portions of its resources, command, or both, and continue to be operationally viable. This is due to the ability of the individual tactical units to readjust their strategy based on their capabilities.

Sociological Limits

In all three GW models presented so far, a limitation on military actions exists, which its parent society, as a nation-state, imposes. This limitation precludes the military and government, at least openly, from engaging in acts that its society considers morally reprehensible. Thus, one country may allow torture under certain conditions, while another will not under any circumstances. Likewise, one country will not only permit, but also demand, that a passenger airliner be shot down with foreknowledge that it will hit one of its major population centers and cause a catastrophic loss of life, while another will refuse to do so.

These standards change as the social factors of a country change, but provide the framework under which legal precedent is established; and impose limits on what will be permitted by a nation's military on its behalf. Genocide, rape, and looting are but a few warfare tactics that collectively have come to be identified as "war crimes," and invoke an international response. Bribery, murder, and other more individual level crimes can invoke cultural outrage and ensuing legal prosecution, depending on the socio-logical disposition of the nation-state the individual represents, or is opera-ting in.

Time limitations are also placed on an actual war, due to economic factors. When the nation-state is dependent upon international trade, a long-term war can have crippling economic effects. Indirect social impacts, such as inflation, can erode the electorate's confidence in the government. Also, few nation-states will engage in warfare to the point of their total destruc-tion, favoring instead some form of compromise embodied in a structured surrender. The question then is how far a nation-state is really willing to go. In contrast, practitioners of fourth-generation warfare (4GW) do not have any such limitations, because they do not represent a nation-state.

FOURTH-GENERATION WARFARE

Fourth-generation warfare (4GW) began with Mao in 1924 (Hammes, 2004) and is considered a "total war" model, using every conduit imaginable as a way to fight. The reason for 4GW association with Mao is because of his use of the rural peasantry which has its basis more on matriarchal/patriarchal, clan/tribal affiliations than a national identity. Mao's model broke ranks with the conventional belief that only the middle class could induce a revolution. 4GW's very nature lacks limitations as it ignores the nation-state restrictions on warfare, particularly time. Its full definition is still under development due to its transient nature. (This is normal in GW theory as only in retrospect does a generation become defined.) Ongoing evolu-tions make a comprehensive theory difficult to define, but a few principal traits have emerged.

First, 4GW is not insurgency, as that would imply actions taken exclusi-vely within a nation-state's physical boundary, with only indigenous persons involved in the insurgency. Combatants engaged in this type of warfare do not align themselves from a nationalist perspective, but a religious, social, political, clan/tribal, or economic one. Age, gender, sexual orientation, and other personal factors of the combatants are not necessarily a limitation for 4GW practitioners, as they are with nation-states that must balance their social values with military necessity. 4GW is not restricted to the centuries-old guerrilla warfare tactics from which it originated. Rather, it is the offspring of guerrilla tactics mated to the open-access/off-the-shelf technological developments of modern times. Alliances are based on situations where both parties are moving in the same direction, toward whatever objective they both seek. When the direction changes the alliance ends, and the two parties proceed independently. The result is a transitory battlefield that is based more in perception than traditional geographical parameters.

The easiest way to understand the 4GW concept is to imagine the world before the Internet. An individual seeking an alliance for social change would spend a long time just seeking out like-minded persons interested in a particular issue. Security was a prime concern, as certain issues would invariably alert law enforcement and draw surveillance. Furthermore, activities by such persons or small groups were geographically isolated and generally remained fixed on a single issue. These individuals and small groups did not pose much of a threat to a nation unless their platform was broad enough to garner some degree of popular support. With the Internet that same individual now has the ability not only to find like-minded persons, but also to coordinate with organizations that are not like-minded, but share similar goals. The result is the development of enhanced capability from force multiplication through networking on a global scale. This scope of cooperation is far out of proportion to the traditional concepts of guerilla warfare that had been geographically fixed.

For example, 4GW practitioners may kidnap or kill several CEOs of large corporations around the world that have operations in Brazil, giving as a rationale their desire to end these corporations' exploitation of the rain forest. The replacement CEOs and board members would probably suspend those operations for their own safety, at least until the corporation could assure their security. A series of similar attacks over time would seriously erode international market confidence in dealing with Brazil. The result would be a weakening of the Brazilian economy that in turn could destabilize the government. This is in sharp contrast to a traditional insurgency that would have attacked the corporate operations in the field. Pressure from the CEOs in those situations would have called for the host government to provide increased military protection for their operations without them being halted.

For a 4GW practitioner the boundaries of social norms that restrain a nation-state from open use of coercive tactics do not exist. Instead, they are used as a benchmark to determine what level of force or action may be necessary to achieve the desired results. The further the action is from those social norms, the more effective these norms become, as it forces the nation-state involved to face issues that demand action, but require it to be proportional to existing social restrictions. Hence, the massacre of hundreds of persons in a mass transit attack may only result in a lifetime confinement under the nation's criminal justice system, while the impacted population perceives that a more just response would be the death penalty. The difference can cause fissures within the social fabric that can lead to new openings for the further exploitation of 4GW.

For 4GW, conflict time is measured in decades, as opposed to the weeks, months, or years that nation-states are restricted due to social and economic factors. From a 4GW perspective, a protracted engagement is preferred, because it uses those social and economic factors that restrict nation-states to a short duration, and is therefore a strategic weapon. The old adage that "a guerilla warfare practitioner wins just by staying alive" is at the core of this. The longer the conflict can be drawn out, the more likely an economic or social concession will be made due to the shifting political alliances inside a nation-state. The result again is a weakening of the nation-state's social fabric for further exploitation.

The most persistent characteristic of 4GW is the use of any opportunity to inflict damage. This operationally means that the difference between military and civilian targets is nonexistent. Attack modality can be anything from events causing mass casualties to destroying the main computer files of a large corporation causing economic hardship or business failure. Limitations only exist in the context of the practitioner's objective (e.g. the decision by the Irish Republican Army (IRA) to limit civilian casualties) or technological capability. Opportunity for reciprocity is denied for two reasons. The first is the general inability to identify the supportive structure of the 4GW practitioner that would provide parity with their attack. In reality, those who do support the 4GW practitioner may also be the nation-state's allies on another level, fostering a political conundrum. The second is the moral and social barriers of the nation-state that are its foundation. For example, a 4GW participant would not think twice about using torture as a tool in accomplishing their goals. An agent for a nation-state may also do so, but with the knowledge that his own society will condemn him for doing so, and that he may face criminal proceedings if his actions are discovered. If a nation-state elects not to condemn the illegal actions of its agents in so-called peacetime, its social fabric is weakened for other potential abuses.

The traditional premise of this form of warfare, which is rooted in insurgency or guerrilla warfare, has been to weaken or eliminate the bonds between a government or occupying force and the population. These are not a series of moves or actions that pit the nation-state's strength against itself as much as they are a concerted effort to bring its very legitimacy into question, and force its populations to grapple with issues that define who they are and what they stand for. By creating an event that forces a government to respond invariably with insufficient or excessive force, a tightrope is created in trying to balance security measures against the underlying values of the society. If the government becomes too proactive, it risks alienating portions of its society. If it does not become proactive enough, other portions of its society are left with the impression of vulnerability and isolation. The result in either case is greater opportunity for exploiting the gaps created.

The attacks are kept to short duration due to the need to insure positive control over tempo. Protracted operations run too great a risk of losing control due to the inevitable response such an event would draw, and the inability to call on reinforcements. Surprise, speed, and violence of action must be compressed into as tight a presentation as possible in order to provide every assurance of success. The goal is to raise the cost of a prolonged struggle to such a level that the nation-state targeted becomes politically compromised from pressure through economic, social, religious, or other means. This situation is exacerbated by the changing roles of private sector organizations on the world stage.

Private sector organizations of the industrial age relied on hard technology and raw material resources that were geographically fixed, placing them under the ultimate control of the nation-state. Sale and exportation of the products, as well as the facilities and raw materials, could be seized, taxed, and regulated. Current private sector organizations are transitioning toward intellectual technology or property that is much more difficult to control and is not as dependent upon raw materials or specific locations. The result is the ability of these organizations to shift operations from country to country based on economic factors. This ability is eroding the influence and/or control of nation-states as it makes the market more fluid, resulting is faster economic changes. For example, imagine for a moment that it is possible for the software company Microsoft to move its entire operation to India after an unfavorable American lawsuit. Once incremental costs in one country exceed a certain threshold, established by the business opportunity or viability of other countries, there is little to hold these corporations in place.

Businesses rely on nation-states for services such as litigation, provision of currency standards, and trade protection, which can be obtained from a variety of nations. The mobility of private sector revenue generators may force nation-states to bend their internal rules and international relations to keep these private sector economic engines. As government policies favor corporations, the balance of power between government and business may change. A multinational corporation may prop up a friendly but tyrannical and undemocratic government, while exploiting the population. Because government no longer protects their interests, certain segments of the population may engage in 4GW against their own governing authority.

Limitations of Fourth-Generation Warfare

There are limitations to 4GW as there are for 1–3GW. It will never become a mechanism of nation-states because it violates the norms of both internal and international relations. Its employment would in effect undo the conventions nation-states were built upon since the Peace of Westphalia. Elements such as the Chinese Communists, who have used it in the past to incite the population and/or destabilize the existing regime, have had to transition to one of the previous GW models in order to complete the transition to nation-state legitimacy (Galula, 1964). 4GW is purely an offensive system whose only defense is anonymity. If that is lost, their ability to leech off the infrastructure of nation-states is jeopardized, as is their control of operational tempo. If identified and fixed in a geographical location, a 4GW combatant cannot survive, because a 1GW-organized nation-state can bring sufficient forces to bear and eliminate the threat.

Therefore, a time restriction exists for 4GW practitioners to conduct their operation(s) with a reasonable chance for success. That includes the intelligence and logistics phases, as they are prone to compromise. Going beyond a certain time frame increases the likelihood of detection, resulting not only in compromising the mission but also in loss of anonymity. Currently, complex missions necessitating multidisciplinary capability, real-time coordination, and/or the ability to improvise similar to 3GW have been avoided in favor of missions that are more akin to offensive set-paced 1–2GW operations. Therein lies the vulnerability of 4GW: if first responders become more interconnected and situationally aware, their reaction times to response, organization, and eventual recovery decrease. This will in turn place additional emphasis on pre-mission planning for 4GW practitioners, resulting in either simpler attacks or longer periods of inaction due to the necessity for increased intelligence gathering in the pre-mission cycle to maintain the current level of complexity. Still, an enhanced civilian response capability will mitigate the impacts of 4GW attacks, and marginalize the perceived success, by denying the visual impact of chaos that fuels the perception of government ineptness that 4GW practitioners seek to attain.

In keeping with that thought of perception, how are success and failure in the war on terrorism even determined? The answer lies with the population, as it is a subjective issue. However, there are a number of mechanisms that are used to judge first responder effectiveness that could be used by Public Information Officers (PIOs) in accounting for the efficiency of the response effort when addressing the media. Casualty rates, numbers saved versus dead, collateral impacts, and injury rates among law enforcement, fire, and medical personnel are traditional areas addressed, but can be misleading. How many died instantaneously? Were law enforcement, fire, and medical personnel injured as part of the incident or while responding to it? While these factual issues may appear trivial, they can cast doubt on the effectiveness of a government.

Concentration on performance through time carries a different significance, as it establishes quantifiable standards. How long did it take for responders to arrive on scene, to establish on scene command, till the first or last victim was dispatched to the hospital. How long was the time to clearance of the scene? There are but a few of the criteria measured in quantifiable terms. They provide a better explanation of the effectiveness of the response system and provide a context of comparison not only to similar events, but also to normal response occurrences.

First Responders

It has been said that a country's military is a reflection of the society which it serves. Law enforcement and fire are also a reflection of the value system, customs, and traditions of their parent country. Because of this similarity, and their clear hierarchical command structures, a superficial resemblance exists between military and civilian responders. However, each plays a distinctly different role in the maintenance and preservation of the society they serve, and have distinct perspectives on their purpose and function. Even among first responders there are subgroups that specialize in particular areas. Law enforcement activities are not the same as firefighting or medical activities. Employment of one in place of another is not comparable to substitution of naval resources for army mission assignments. Each civilian first responder organization is uniquely different as it serves a small portion of the community it belongs to. In some cases these first responder organizations work separately due to tradition, politics, and social orientation. In cases of other communities these same organizations will work at such an integrated level that it is difficult to determine, for a casual observer, who is in charge. Additionally, initial impressions of individual organizations can be misleading.

Upon initial examination law enforcement appears to be very rigidly structured, with hierarchical command structure borrowing a great deal from the military. Upon closer examination you find a very high degree of individual discretionary authority vested in the field levels. Senior personnel do not command so much as coordinate and work to build consensus with their subordinates. Fire is another hierarchical command, but appears superficially more lax in structure. In reality, firefighters are more rigidly structured than law enforcement because all fire operations are team events, as a result placing demands on its members to protect one another and fulfill a specific function, utilizing equipment and tactics that demand precise coordination.

If you placed a full backpack on top of a drum filled with HAZMAT blocking the door to an emergency room and had a representative from fire, law enforcement, and medical service assess the situation, you would get three entirely different answers. On the other hand, if you had three people from any one of the professions perform the evaluation, their answers would probably be similar. The unique response of each civilian service is due to the specialized nature of each discipline's area of responsibility. It is an operational problem when dealing with terrorism because the very first responding entity to an event is likely to be composed of representatives of these three disciplines. They will provide the initial size-up of a complex major terrorist event, a role that is well beyond their normal scope of work. While the actual tasks may be beyond their capability individually, they have the possibility to play a significant positive role in the eventual outcome of the event if they work collectively with their counterparts. Like the proverbial three blind men trying to describe an elephant, each touching a separate part, each service will develop its unique perspective. However, by sharing the observations, their collective knowledge and decisions could result in establishing the initial on-scene organizational framework for other units to begin filling immediately upon arrival. A single entity providing the evaluation might miss key factors, and require the Incident Commander to obtain a reassessment and redeployment of resources. To facilitate that capability, various entities from fire, law enforcement, and medical service must be able to work in ad hoc units to provide mutual protection and initial coordination of their respective chains of command.

Some assume this to be how first responders function currently. In reality, these elements rarely associate with persons outside their own discipline, and will defer to their supervisors in situations beyond the norm. In essence, specialization has resulted in very clear lines of responsibility between the various response organizations, leading to clear command responsibility. For example, at a house fire the Incident Command clearly belongs to the fire department, while law enforcement supports by directing traffic and controlling crowds, and medical personnel support by caring for rescued victims injured by the fire. At an ongoing crime law enforcement is Incident Commander, with fire and medical personnel supporting injured victims, as law enforcement mitigates the criminal activities. The problem occurs when several different areas of responsibility overlap and present conflicts with long-established priorities divided between the various responding disciplines. A secondary device found after medical assistance is rendered to a nonmovable victim could be such an occurrence. Law would order the area cleared, while medical personnel would balk at leaving a patient, and the fire department would want to continue the rescue attempt. While such scenarios are clearly situationally specific, understanding the underpinning responsibilities of each discipline involved reduces operational conflicts and friction, and decreases the time needed to develop an appropriate response. Therefore, unless extensive planning and training (both facilitating understanding of the other disciplines) have preceded the event, there may be conflicting evaluations of the event, leading to conflict over scene control and priorities.

While the above situation casts a gloomy picture, there are two areas that all first responders agree upon: the first 20–30 minutes of an event are the most critical, and injury or death of a first responder equals failure. In the case of the former, an event handled properly at the onset will progress in a manageable, if not controllable, fashion. This requires a proper size-up of the situation, establishment of positive control over the impacted area, and setting up of a command system to manage resources. If these areas are ignored, the event will unfold of its own accord, possibly with responders and citizens contributing to the problem instead of solving it. This includes injury and loss of life among first responders.

For first responders, injuries and death equate to failure. It violates their first priority: preserve life. There is no middle ground on this issue, as loss or injury of a single first responder can have a dramatic impact upon not only the organization's operational capability, but also on its social framework. First responders spend their careers with the same personnel and develop lifelong bonds, not only between themselves, but also among their families. In some instances, an organization will actually identify itself as an extended family. These organizations are also run very leanly, with little or no overstaffing, and in some cases are only able to fulfill all job requirements by staffing with mandated overtime. This puts the safety of trained personnel in a more visible and essential position than in other organizations with more personnel resources.

Trying to integrate other organizations, including the military, into a response to a terrorist attack on a civilian target becomes a concern at this point. The underpinning values that enable consensus among first responders (first 20–30 minutes, no injuries or loss of life) may need to be emphasized in order to assure continuity of focus. This is no trivial matter, as other civilian organizations are rarely placed in harm's way in the same manner as first responders are, and therefore might not be attuned to these issues. Furthermore, the military has the opposite view in this regard. Their principal area of responsibility necessitates placing the life of a military responder at a lower priority level than the nation's security. In a high stress civil authority environment, such as a 4GW event, when personnel are prone to fall back instinctively to their training, this issue, if left unaddressed, can have unintended consequences and result in friction between first responders and various other entities present. This becomes less of an issue as an event unfolds due to delays in response for military, non-sworn, and social service agencies.

Proposal

An opportunity exists in using the two areas of agreement (first 20–30 minutes and preservation of life) as common ground toward building consensus on response among first responders. Consider for a moment the use of the 3GW concept of center of gravity (*schwerpunkt*), with the initial period and preservation of life as the main focus. This would provide a standard base of reference between first responders confronted with a catastrophic event, and enable rudimentary coordination and force protection in areas normally addressed individually by the specific disciplines present. Fire, for example, would advise law enforcement on HAZMAT present, while law enforcement would call fire's attention to tactically suspicious situations. Coordination at this immediate level requires cross-disciplinary trust, consensus on the primary goals, and a team—not competitive—approach to evaluation, planning, and execution of response. While this sounds like a simple concept, implementation can be an issue, as existing organizational structures could perceive this high level of coordination to be an infringement on their sovereignty as each entity has to sacrifice some control to obtain coordination. The matter would need to be addressed as a safety concern for all first responders in order to avoid those sorts of conflicts from occurring. Building upon mutual respect, trust, and cooperation would strengthen the response capability for all who participate during the initial phase of response to a 4GW attack.

Daily interaction among different responding disciplines should be encouraged. Incorporation of an exchange program, similar to military observer programs, among civilian disciplines should be institutionalized. This could include having staff at key points in their careers participate in activities two or three levels above their current positions, and senior personnel responding to small events as evaluators. All of these would strengthen the bonds in a chain of command, provide consistency in understanding, and enable cross-linking with other organizations. However, beyond a certain level, a framework of reference to determine how these various entities can function with one another for a protracted period is needed.

As noted above, an organization's structure is not necessarily the best reference point to determine capability. Therefore, it should be the functional components, or how they actually work, that should be assessed. In that light, the traits of 1–3GW could be used to "type" what functional level a particular organization has attained. This would indicate what could be reasonably expected of it, and identify the operational barriers that may exist for it to be compatible with other organizations that share the same arena of responsibility, whether spatial or jurisdictional, at an event.

Such a typing would enable optimization within each of the generational models, and encourage development toward 3GW, when possible. At the very least, it would minimize waste in attempting to replicate features (e.g. equipment, training, and tactics) that are outside the functional capability of an organization. The acquisition of redundant or incompatible equipment occurs frequently, particularly post 9/11, with trailers and storage rooms of equipment, but no clear concept on how to employ it. All too often this is due to a department trying to replicate ability without determining if their functional capability can support it, or even if a legitimate need exists.

While this initial response and coordination would be primarily to ensure the safety of the responders, it would also assist in the transition to the next phase. Once safety and security have been established and linkage between the various on-scene assets has occurred, a more formal and permanent structure will be needed to maximize the resources available. An Incident Command System (ICS), or similar hierarchical structure, that relies on a span of control will be required to enable the incorporation of assets that are functionally 1 and 2GW. These agencies may require assistance in understanding the situation and adapting while being assigned to work with, or in direct support of, other assets with differing functional levels. For instance, a prisoner work detail (1GW) may be assigned to support a HAZMAT team (3GW) on mitigation efforts in a cold zone. While it is obvious that the HAZMAT team would work through the guards in directing the prisoner's actions, the reciprocal issue of prisoners' identifying potential hazards is raised. Who do they address, the guards or the HAZMAT personnel? It is these sorts of functional issues that are the responsibility of staff, above both groups organizationally, to identify and address if they cannot do so between themselves. While the above example is an extreme case from both a real-world and functional perspective, it makes the issue clear, particularly when agencies that are organizationally identical but functionally different are put together and discovered to be incompatible.

Typically, an ICS structure with many agencies will take on the functional attributes of 1GW or 2GW, due to the need to address the lowest common reference points between agencies. However, this need not be the case, and should be avoided, as it prevents optimizing resources, particularly of 3GW. By understanding and typing the functional level of each asset it is possible to span out resources in a way that would not hamper the activities of those organizations that are functionally more developed. The solution to this issue is through cross-exposure, by conducting training and exercises between the various potential responders to an event. This is not to say that every agency would need to participate fully. As noted with the prisoner/ HAZMAT example above, the exercise or training may only need to involve key personnel, particularly with 1GW functional agencies that are envisioned to respond after the safety and security portions of an event have been addressed.

By conducting training and exercises that involve different disciplines, a common understanding of each organization's functional capability, as well as their responsibility, is developed. Gaps between agencies can therefore be identified in the process. While not all of the gaps will be successfully addressed, that they are "seen" enables participants to mitigate their potential impact at the time of an event.

Conclusion

The world is a dangerous place and is growing evermore so as our technology progresses at an unprecedented rate. There will always be those ready to exploit the latest advancements for illicit purposes. The demands on first responders are likewise increasing because of those advances, while at the same time they are forced to do more with fewer assets.

The question, then, is how prepared is the society to absorb the impact, and continue, when experiencing sudden multi-casualty activities? This is more than an academic question and can be quantified by the value placed by that society on the institutions that form the bulwark of its defenses, both externally and internally. Coordination among those institutions, as well as their counterparts in neighboring friendly countries, will become increasingly important as social, economic, political, and religious factors become more complex. A nation-state can ill afford the loss of the confidence of its citizens, as that is the ultimate objective of 4GW practitioners as we understand them today.

APPENDIX 3: REVIEW OF THE BRAZILIAN EMERGENCY MANAGEMENT AFTER THE RADIOLOGICAL INCIDENT IN GOIANIA IN 1987

The biggest experience managing the impact of large-scale radioactive contamination on society due to loss of control over a radioactive source has been gained from the incident in Goiania, a city in Brazil with approximately one million inhabitants. In September 1987, two thieves removed part of a teletherapy machine from an abandoned clinic for the metal value of the shielding material only. The operators of the clinic had moved to another location, leaving the strong radiation source behind in an unattended building.

Reaction by the Authorities in Goiania

On September 13, 1987, after several attempts to open the radiation source, using brut force and fire, a scrap yard dealer finally succeeded. This resulted in an uncontrolled release of Cs-137 from the opened capsule, which triggered multiple events, ultimately causing several deaths, and hundreds of persons contaminated or suffering from radiation injuries (International Atomic Energy Agency, 1998). However, the radioactive release was not discovered for 16 days. It took until September 29, 1987, for the authorities to finally become involved due to unfortunate circumstances.

Several persons became sick and sought medical treatment. Initially, they were erroneously diagnosed as suffering from dehydration and tropical diseases; in this period five hospitals were contaminated.

The news about the radioactive contamination of Goiania residents spread quickly across the city and its approximately one million inhabitants. In order to regain control the authorities identified those who were most likely to have been contaminated and directed them to the Olympic Stadium for radiation screening and triage. By February 1988, the total number of persons measured exceeded 125,000. Subsequently, the authorities had to provide three different facilities for the 260 persons suspected or confirmed as being contaminated.

After a medical examination, external radiation monitoring, and laboratory analysis of these persons, the most serious 14 cases were transferred to the Marcilio Dias Naval Hospital in Rio de Janeiro, the center for radiation accident casualties. This transfer of victims to Rio de Janeiro resembled a war-type operation, because each individual represented a source of

221

F. L. Edwards and F. Steinhäusler (eds.), NATO and Terrorism, 221–224.
© 2007 Springer.

radiation, potentially contaminating anyone in close proximity. Four of these persons died in October 1987, their whole body dose ranging from 4.5 to 6 Gy.

In order to establish the extent of the contamination of the city and its surrounding areas, search teams were equipped with different types of radiation detectors. They investigated the whole city, including the identification of contaminated cars passing by on the highway. The operators and their detectors had to withstand the high humidity and hot conditions (temperatures above 30°C). In addition, aerial surveys were carried out over 67 km^2. This revealed that in Goiania city seven main areas were contaminated with dose rate values up to 2 Sievert/hour measured at 1 m distance, necessitating two areas to be evacuated.

Consequences for Society

Due to the extended period before the incident was discovered, the radioactive contamination spread to multiple sectors of society:

- 42 houses required decontamination.
- 50,000 rolls of toilet paper (total activity: 0.67 TBq) had become contaminated.
- Some banknotes had a dose rate of 0.8 mGy/hour each.[1]
- 3 buses, 14 cars, and 5 pigs were found to be contaminated.
- Travelers spread the contamination to the nearby townships up to 100 km from Goiania, such as Anapolis, Aparecida, Trindade, and Goias.
- Transfer of contaminated products during trade added to the complex task of delineating the contaminated areas: contaminated paper in São Paulo and highly contaminated lead in Goias Velho were recovered at distances up to 830 km from Goiania.

Authorities set up physical barriers at the borders of the State of Goias to hinder cars from leaving the state, and provided 8,000 residents with an official certificate that they were not contaminated. Members of the public complained about significant deficiencies in the information policy, and the unduly complicated organizational decisions. Altogether, it took the authorities 6 months to decontaminate Goiania and its environs.

---

[1] 1 Milligray (mGy) = 1/1000 Gy.

Attempts to Manage the Crisis

It required altogether 3 days after the discovery of the incident for the professionals of the Brazilian National Nuclear Energy Commission (CNEN) to address the most urgent radiation contamination, public health, and medical needs. Subsequently, the attention of the expert community from CNEN and several other organizations focused on monitoring the population, cleanup operations, and long-term consequences resulting from the incident.

Initially, there was a high level of panic among the residents: a few days after the official disclosure of the *Telephone Hotline* at least 5% of the public presented symptoms of contamination. Therefore, it was necessary to establish reliable methods which allowed the identification of radiation victims. For this purpose the radiometric investigations were supplemented by cytogenetic dosimetry on persons without any radiation syndromes, but who had become potential victims. Injury due to the deposition of radionuclides on skin and mucosa or direct contact with the radiation source was an additional clinical problem.

It is important to note that of the four fatal casualties, three were the results of significant deficiencies in medical management and prophylactic measures, such as poor aseptic environment; multiple evacuations of patients; inadequate antibacterial therapy; and lack of experience of the medical staff.

CNEN staff had extensive theoretical knowledge, but no practical experience with real radiation victims. This expressed itself in frequently observed first shock upon encountering radiation dermatitis. Generally, radiation protection experts were strongly affected by radiation-induced injuries as they were afraid that patients would die. Excessive stress was experienced by overworked experts, who had to deal with the organization of services, procedures, coordination, and the division of tasks, whilst wearing protective clothing and masks for extended periods. In particular, the monitoring of autopsies was found to be unbearable.

Due to the lack of regular trained hospital personnel, the radiation protection experts also had to assume the role of nurses and guards, for example having to cope with patients screaming and—in despair—wanting to escape from the ward by throwing themselves out of the window. This led to psychosomatic disturbances among staff, such as insomnia, gastric problems, and weeping freely when away from patients and coworkers. After 2 weeks the radiation protection team was completely exhausted owing to excess work and emotional stress, with the first group suffering most because of inadequate knowledge of what had happened.

Medical care for radiation victims met several obstacles. The hospital personnel lacked preparation for treating radiation injuries and showed fear, with some of them leaving the victims unattended and going on strike instead. Also, the lack of response from the Goianian medical community is to be noted, despite the appeals made by physicians. Generally, there was a noticeable lack of medical staff, nurses, paramedical workers, and cleaning personnel in the hospital services. Furthermore, the significant deficiency of laboratory support to perform clinical analyses, the inadequate laundry, and the unsuitable system of collecting hospital and decontamination waste added to the difficulties.

APPENDIX 4: REVIEW OF US EMERGENCY RESPONSE
AFTER THE TERROR ATTACKS IN NEW YORK CITY
ON SEPTEMBER 11, 2001

Evacuation

Evacuation is a powerful tool in the arsenal during emergency procedures, aimed at the removal of inhabitants from a place of danger for protective purposes. On September 11, 2001, the Twin Towers of the World Trade Center were on fire, and there was little prospect of getting these fires under control in a timely manner, if at all. While evacuating the employees from these buildings would seem to be a foregone conclusion, there was confusion about what procedures to follow.

In the USA there has never before been a collapse of a high-rise building due to fire. The standard procedure for high-rise operations, based on US Fire Administration training, is to evacuate everyone above the fire floor, and two floors below. Due to the damage to the structures caused by the jet planes, especially to the evacuation stairwells, it was impossible for most of the employees above the fire floors to exit the building. In addition, the standard direction for employees to follow when there was a fire alarm for an unknown cause was to phone the Port Authority Dispatch center for directions on how to proceed. Communications failures along the chain of command left those operators without accurate situational status information, so they could not provide direction to the callers. In the end, people made their own decisions whether to stay or leave, mostly based on their own sense of danger and the direction of their colleagues.

The actions taken at the time of the terror attacks on September 11, 2001, illustrate the difficulties all parties involved were facing in effectively evacuating the people at risk. The following analysis is derived from *The 9/11 Report* by the officially appointed US National Commission (Kean and Hamilton, 2004).

Pre-9/11: Deficient Evacuation Procedures in the Twin Towers

The authorities responsible for emergency evacuation procedures at the Twin Towers had been in the unique situation of having been able to learn from the shortcomings revealed after the previous terror attack on February 26, 1993, when a 680 kg bomb hidden in a van detonated in the parking lot

F. L. Edwards and F. Steinhäusler (eds.), NATO and Terrorism, 225–233.
© 2007 *Springer.*

beneath the towers. At that time the following main shortcomings were discovered:

- Loss of electric power
- Failure of the public address system
- Failure of the emergency lighting system
- Impassable unlit stairwells due to high smoke concentration
- Inability of the radios used by the Fire Department of New York (FDNY) to communicate effectively within the concrete and steel building
- Overwhelmed 9-1-1 emergency call system
- This resulted in an evacuation time for the Twin Towers of more than 4 hours in 1993. Subsequently, US$100 million were spent initially to improve the situation at the Twin Towers (technological, structural, and physical changes; enhancement of the fire safety plan; reorganization of the security staff; establishment of fire command stations in the lobby of each tower; fire drills at least twice a year; installation of an emergency intercom phone and a repeater system to enhance the FDNY radio communications in high-rise environment).

However, despite these improvements, significant deficits continued to exist, such as: (a) during the fire drills occupants were not directed into the stairwells; (b) occupants were not provided with information about the configuration of stairwells, the existence of transfer hallways, the installation of smoke doors; (c) neither full nor partial evacuation drills were held; (d) occupants were never told not to evacuate, and were not informed that rooftop evacuations were not part of the evacuation plan; (e) occupants did not know that the doors to the roof were kept locked. In fact, there was no protocol for rescuing occupants in case they were trapped above a fire in the Twin Towers.

Evacuation on 9/11: Lessons (Un)Learned

NORTH TOWER

Floors 93–99 of the North Tower (1 World Trade Center) were hit by the American Airlines Flight 11 at 8:46. The deputy fire safety director in the lobby was unaware of what exactly had happened until approximately 8:56. Occupants on floors that had generated computerized alarms were initially advised to descend at least two floors below the smoke and fire and wait for further instructions; however, within 10 minutes the safety director began instructing a full evacuation.[1] Unfortunately communication between the occupants in the North Tower and the emergency responders was severely restricted:

- Public-address announcements could not be heard in many locations.
- It is likely that many occupants were unable to use the emergency inter-com phones due to the damaged building infrastructure.
- The 9-1-1 emergency call system was unable to handle the resulting large volume of calls.
- Transfer of 9-1-1 calls to FDNY dispatch operators in Manhattan were often subject to delays.
- Many calls suffered from premature disconnection.
- The operators at the 9-1-1 call centers lacked information about the location of the impact zone and the magnitude of the plane impact.
- FDNY dispatchers also had no information about the location of the impact zone and the magnitude of the plane impact.
- 9-1-1 operators and FDNY dispatchers could not inform occupants whether the fire was below or above them.
- 9-1-1 operators were unaware of the decision by NYPD Aviation that rooftop rescues were impossible, i.e. they could not provide any advice on whether to ascend or descend.
- Neither 9-1-1 operators nor FDNY dispatchers were aware of the decision by the FDNY chiefs in the lobby for immediate evacuation of all occupants.

This lack of adequate information and communication resulted in incoherent advice given by 9-1-1 operators to occupants below and above the fire, i.e. (a) to remain where they were and await the arrival of the emergency personnel; (b) to break windows; (c) to evacuate, if they could.

By 8:57 FDNY chiefs issued an evacuation order for all occupants, but this was not conveyed to FDNY dispatchers or to 9-1-1 operators. The evacuation procedure itself was hampered by confusion due to deviations in the stairwells, apparently locked doors,[2] heavy smoke rising to the upper floors, and isolated fires. Once occupants had reached the street level, some of them were killed by falling debris.

---

[1]This is in contrast to the report that the Port Authority fire safety director had advised full building evacuation within 1 minute of the building being hit.

[2]Actually jammed by debris or shifting of the damaged building.

SOUTH TOWER

In the South Tower (2 World Trade Center) information about the events in the North Tower varied considerably, resulting in largely incoherent advice given by officials and greatly varying reactions by the occupants:

- Some occupants decided on their own to leave.
- Some occupants were advised by fire wardens to do so.
- At about 8:49 tenants were advised over the public-address system that their building was safe and that they should remain in, or return to, their offices; therefore, many occupants remained on their floors or reversed their evacuation and went back up.
- Security officials in the ground floor lobby and in the upper sky lobby provided similar advice.
- Port Authority Police advised some callers to stand by for further instructions, whilst it strongly advised others to leave.
- At 9:02 civilians were advised over the public-address system that they could begin an orderly evacuation, if conditions warranted.

At 9:03 United Airlines Flight 175 hit the South Tower and crashed through the 77th to the 85th floor. In contrast to the North Tower, one of the stairwells was likely to have remained passable from top to the lobby. Several people used this stairwell to descend. Attempts by people—already waiting to evacuate as a consequence of the attack on the North Tower—to squeeze into packed express elevators failed. Many remaining alive in the impact zone above the 78th floor ascended the stairs. Attempting to reach the roof they met locked doors. The "lock release" order transmitted at about 9:30 through the computerized security system could not be executed due to damage to the software controlling the system, so the doors to the roof remained locked.

The 9-1-1 operators had similarly inaccurate information, which led to inadequacies in handling this situation. Survivors calling 9-1-1 for directions, location of the fire in relation to their own position, or indicating that they were running out of oxygen were advised to remain where they were and wait for assistance. It is likely that many occupants did not attempt to evacuate for this reason, although it is highly probable that one stairwell may have been passable all the way.[3] There is contradictory information about whether the public-address system remained functioning after the plane had hit the building.

---

[3] By 9:59 at least one person had descended from the 91st floor on a reportedly almost empty stairwell.

The First 17 minutes for the First Responders: Lack of Information and Communication

Although the FDNY response started within 5 seconds after the plane had hit the North Tower, it took another 14 minutes before the Chief of Department arrived at about 9:00. It required almost 74 minutes for 22 of the 32 senior chiefs and commissioners to arrive at the Twin Towers. Operations themselves started earlier: the highest-ranking FDNY officer, a battalion chief, two ladder, and two engine companies arrived about 6 minutes after the crash; two scouting units began climbing a stairwell 1 minute later.

There was no information available on the following topics: (a) what the impact floors were; (b) whether any stairwells into the impact zone were clear; (c) whether water would be available for fighting the fire on the upper floors; (d) what the fire and impact zone looked like from the outside; (e) the explosion had blown out elevators and windows in the lobby; (f) the dire conditions which caused occupants to jump from the building.

The Port Authority Police Department (PAPD) Chief of Department and Superintendent arrived at 9:00. However, the response by PAPD was uncoordinated for two reasons: (a) it lacked written standard operating procedures for personnel responding from outside commands to the Twin Towers during such a major incident; (b) officers from some commands were using radio frequencies, which were not interoperable. Communication between PAPD and staff at the South Tower was also compromised, i.e. the evacuation order given by the PAPD commanding officer at the Twin Towers at 9:00 could not be heard by the deputy fire safety director in the South Tower because it was given over the World Trade Center police channel W.

New York Police Department (NYPD) sent approximately 22 lieutenants, 100 sergeants, and 800 police officers from all over the city to the Twin Towers 12 minutes after the hit. However, many of them were diverted on the way, since they had to intervene in emergencies which were related to the terror attack. NYPD coordinated with PAPD in the closing of tunnels and bridges leading into Manhattan.

Four minutes after the hit the NYPD Aviation Unit dispatched two helicopters to the Twin Towers to assess whether special rescue operations were feasible. Already 12 minutes after the building was hit, a helicopter pilot determined that they could not land on the roof because of flames and smoke. Neither FDNY nor 9-1-1 operators were informed that rooftop rescues were not being undertaken.

The Office of Emergency Management (OEM) of New York City contacted FDNY, NYPD Department of Health, FEMA and the Greater Hospital Association 2 minutes after the hit. Already 2 minutes later a senior representative of OEM arrived in the lobby area of the North Tower and acted as OEM field responder.

In total, over 1,000 first responders had been deployed within the first 17 minutes after the first plane had hit the North Tower. The subsequent period was characterized by the arrival of excess FDNY personnel, ranging from elite units to self-dispatched units, staff members in administrative positions, and off-duty personnel (violating the instructions not to join).

The Time Before the Collapse of the Twin Towers

REPEATER SYSTEM FAILURE

Once both towers had been hit, FDNY chiefs in the North Tower wanted to activate the repeater system. One of the two buttons on the console had already been pressed at 8:54 in the North Tower, enabling communication between FDNY portable radios on the repeater channel, and listening to these communications over the repeater's master handset. However, since nobody pressed the second button on September 11, 2001, transmitting on this handset was impossible. Also, volume-settings on the console may have been set too low to hear communication attempts by another chief over a portable radio. Altogether, FDNY chiefs in the North Tower lobby decided not to use the repeater system. On the contrary, firefighters in the South Tower used repeater channel 7.

INADEQUATE SITUATION AWARENESS

Situation awareness among the FDNY chiefs in the North Tower was deficient:

- FDNY units climbing up the staircase had not been advised otherwise and communicated on tactical channel 1. As they climbed higher, their ability to communicate with chiefs became sporadic due to the limited effectiveness of their radios in high-rises and communication overload on this channel.
- Due to the lack of critical information there was disagreement whether rescue operations at or above the impact zone were useful, and whether the fire could be extinguished on a limited scale to provide exit routes.
- The assignment of responding units could not be tracked sufficiently well with the magnetic board used.
- A chief in the lobby was unable to reach FDNY dispatch by radio or phone to inquire about rooftop rescue as an option.

In the South Tower there was a significant deficit of FDNY units until 9:30, since units en route to, or already in, the North Tower were not redirected once the South Tower had been hit as well. This caused frustration among the senior chief, battalion chief, and OEM field responder who had started operations in the lobby. They were the only ones using the functioning repeater channel 7, whilst the vast majority of units entering the South Tower did not communicate on the repeater channel.

The operation at the South Tower was hampered by the *disorientation* of some FDNY members: (a) some firefighters mistook the North Tower for the South Tower; (b) some units were unable to find the staging area for the South Tower; (c) 14 units tried to find safe access to the South Tower and ended up in the Marriott Hotel; (d) a chief at the overall outdoor command post also ended up in the Marriot Hotel by mistake and—unknown to him—misdirected units to the North Tower, thinking it was the South Tower.

Collapsing Towers: Unawareness and Misunderstanding

Early in the sequence of events, senior FDNY chiefs had advised that the possibility of a total collapse of the building need not be considered. Therefore, nobody was prepared for the sudden collapse of both towers.[4]

COLLAPSE OF THE SOUTH TOWER

Although the South Tower was hit later, it collapsed first within 10 seconds at 9:59, damaging also the Marriott hotel. Due to this the collapse operations ceased at:

- FDNY overall command post
- Posts in the North Tower lobby
- Marriott lobby
- Staging areas on West Street
- EMS staging areas

The collapse of the South Tower occurred either unnoticed, or was misinterpreted by the FDNY personnel in the North Tower, for example, as a bomb explosion or partial collapse of the upper floors of the North Tower. Although the collapse was communicated at once by an FDNY boat on the

---

[4]One senior FDNY chief, knowing already that the South Tower had indeed collapsed, was of the opinion that the North Tower would not collapse because it had not been hit on a corner like the South Tower.

Hudson River and by the NYPD Aviation Unit, nobody at the North Tower received this information.

This lack of information made the evacuation of the North Tower problematic.

At 10:01 the first evacuation order came from a chief in the process of evacuating the North Tower, himself unaware that the South Tower had collapsed. At about 10:15 the FDNY Chief of Department issued a radio order for all units to evacuate the North Tower, repeating it about five times.

However, several FDNY personnel were unable to receive the evacuation order because of: (a) difficult communication in high-rises; (b) information overload on tactical channel 1; (c) off-duty firefighters who were not equipped with radios; (d) firefighters dispatched to the South Tower who were most likely on a different tactical channel assigned to the South Tower.

In the North Tower FDNY personnel who had received the evacuation order, responded in a nonuniform manner,[5] ranging from delaying or stopping their evacuation in order to assist occupants, attempting to regroup, resting on floors, or simply refusing to take evacuation orders from NYPD officers. Some were even determined not to leave the building altogether as long as other FDNY personnel were inside the building and reascended once they had reached the lobby. As no chiefs were present in the lobby anymore, some firefighters waited in the lobby for over a minute as they were uncertain of what to do—when the building above them started its collapse.

PAPD officers without World Trade Center radios did not receive an evacuation order by radio. Some officers evacuated, based on their own decision or upon consultation with other first responders.

The potential collapse of the North Tower was foreseen, appropriate warnings issued but not heeded: (a) at 10:04 NYPD Aviation Unit warned that the top 15 stories of the North Tower might collapse; (b) at 10:08 a helicopter pilot warned that the North Tower might not last much longer. At 10:28 the North Tower collapsed.

Searching for Victims: Duplication

There was no coordination between FDNY, NYPD, and PAPD about searching the World Trade Center complex for victims. This led to redundant search operations of areas, causing a large number of first responders to be

---

[5]Evacuation orders had not followed the protocol in case of an imminent building collapse, i.e. repeating "Mayday."

in areas of elevated risk unnecessarily. For example, at about 10:05 FDNY personnel at the eastern side of the North Tower lobby, near or actually in the mall concourse, searched the PATH station below the concourse, not knowing that the PAPD had already cleared the area by 9:19.

APPENDIX 5: REVIEW OF US EMERGENCY MANAGEMENT IN NEW ORLEANS AFTER HURRICANE KATRINA, AUGUST 29–SEPTEMBER 2, 2005

Hurricane Katrina represents the first truly catastrophic disaster in the USA since the creation of the Department of Homeland Security. In 2004 Florida experienced a series of four major storms during hurricane season, but each was managed using the enhanced plans developed after the failures in response to Hurricane Andrew in 1991 (Fechter, 2004). Local Florida officials had plans in place, and the FEMA regional office was ready. In 2005, storm predictions energized the states of Florida, Alabama, and Mississippi to move their at-risk populations inland in advance of the storm. In Louisiana past history suggested that the hurricane was unlikely to make landfall in New Orleans, and that if it did, previously used shelters like the Superdome would be adequate for the needs of those who chose not to evacuate. Events proved otherwise.

Early Warning Information Underused by Government and the Public

The emergency response to Hurricane Katrina by dedicated emergency responders provides an insight into the limits of response capabilities to a catastrophic threat. Hurricanes follow an unpredictable course, and although landfall is predicted for up to 72 hours in advance, the exact location makes organizing an adequate response challenging. Local government in Louisiana and the residents of the state did not take aggressive action, although the following information had been available *prior* to the onset of the catastrophe:

- Approximate time of occurrence, resulting in approximately 72 hours of advance warning and time for preparation, which resulted in only 80% of the population
- Geographical area anticipated to be subject to damage, indicating that New Orleans could be at risk
- Mode of damage to be inflicted, forecasting that the levee system of the city of New Orleans may fail due to the impact of storm surge
- Data from previous exercises and modeling, indicating the vulnerability of the flood control system of the area

The summary of events below is based on media releases by the White House, local, national, and federal agencies, as well as reports by electronic

F. L. Edwards and F. Steinhäusler (eds.), NATO and Terrorism, 235–241.
© 2007 Springer.

and print media covering the events prior to, and immediately after, the landfall of Hurricane Katrina in August 2005.[1]

Prewarning Phase August 25–28, 2005: Missed Opportunities

On August 25, 2005, Hurricane Katrina hit Southern Florida, 2 hours after being declared a category 1 hurricane. This caused heavy flooding, resulting in at least seven dead. As the hurricane intensified to a category 3 and headed toward the Gulf Coast, Governor Kathleen Blanco of Louisiana and Governor Haley Barbour of Mississippi declared a State of Emergency on August 26, 2005, and requested a Presidential Disaster Declaration. When Katrina was upgraded to a category 3 hurricane at 5 a.m. on August 27, the President sent a message that a state of emergency existed in Louisiana, and ordered federal assistance to complement the local and state resources. The Department of Homeland Security, and specifically FEMA, has full authority, under the Stafford Act and the National Response Plan, to identify, mobilize, and provide at its discretion, equipment and resources necessary to alleviate the impacts of the emergency. FEMA staff members were sent to Baton Rouge, New Orleans, and Jackson, Mississippi, to coordinate with state officials, and water was stockpiled outside the coastal area. Federal resources, including Urban Search and Rescue Teams and Disaster Medical Assistance Teams across the country, were put on alert. At 4 p.m., Governor Blanco initiated contraflow for all interstate highways to make room for the evacuating population, and at 5 p.m., Mayor Ray Nagin of New Orleans issued a voluntary evacuation order, which was followed by 80% of the residents. At 11 p.m., Dr. Max Mayfield of the National Hurricane Warning Center notified FEMA and local officials that the storm was heading westward to an area that included New Orleans (Talking Points Memo, 2005).

On August 28 at 1 a.m. Hurricane Katrina was upgraded to a category 4 hurricane and further upgraded to a category 5 hurricane at 7 a.m.[2] Media reports indicate forecasters feared that storm-driven waters will lap over the New Orleans levees when Hurricane Katrina was expected to push past the Crescent City on August 29. At 9:30 a.m. Mayor Nagin issued the first-ever

---

[1]Sources consulted for this section include The White House, The Associated Press, Office of the Governor Blanco, Department of Defense, CNN, *Lafayette Daily Advertiser*, *Times-Picayune*, National Weather Service, *St. Petersburg Times*, NBC's Today Show, FEMA news release, Associated Press, Meet the Press (September 4, 2005), WWL-TV, *Chicago Tribune*, *Los Angeles Times*, *New York Times*, Reuters, *Washington Post*, *Fox News*, and *The Guardian*.

[2]The hurricane covered an area equal to the size of the UK; the "eye" of the hurricane had a diameter of about 100 km.

mandatory evacuation order for New Orleans. The President declared a state of emergency for Mississippi and Alabama for preparedness, and declared Florida a federal disaster for the damage sustained from the initial landfall.

At 4 p.m. the National Weather Service issued a special hurricane warning:

> In the event of a category 4 or 5 hit, most of the area will be uninhabitable for weeks, perhaps longer. ... At least one-half of well-constructed homes will have roof and wall failure. All gabled roofs will fail, leaving those homes severely damaged or destroyed. ... Power outages will last for weeks. ... Water shortages will make human suffering incredible by modern standards.

In the afternoon of the same day, Dr. Max Mayfield, director of the National Hurricane Center, briefed President G.W. Bush, FEMA Director Michael Brown, and Department of Homeland Security (DHS) Secretary Michael Chertoff, and warned of levee failure. Late evening first reports came in about waves crashing atop the exercise path on the Lake Pontchartrain levee in Kenner. Subsequently, approximately 30,000 evacuees gathered at the New Orleans Superdome, equipped with roughly 36 hours worth of food.

Impact August 29, 2005: Misinterpretation of Urgency

At 5 a.m. Hurricane Katrina arrived in Louisiana with winds up to 243 km/hour and passed into Mississippi and Alabama. At 8 a.m. the storm surge sent water over the levee at the Industrial Canal. The Army Corps of engineers reported that a barge had broken loose and broken a floodwall, worsening flooding in the Ninth Ward and Arabi, and by 9 a.m. there was 6 feet of water in the community. The President issued disaster declarations for Louisiana, Alabama, and Mississippi, freeing federal disaster response funds. Mayor Nagin reported that water was flowing over the levee and that in the lower ninth ward one of the pumping stations stopped operating. Later that morning a large section of the vital 17th Street Canal levee, where it connects to the "hurricane-proof" Old Hammond Highway bridge, gave way in Bucktown. At 11:30 a.m. FEMA's Michael Brown requested that DHS dispatch 1,000 employees to the region, giving them 2 days to arrive. The storm tore a hole in the Superdome roof where 10,000 evacuees had gone for shelter. By the end of the day the American Red Cross had announced its mobilization to the area. That evening the national Urban Search and Rescue teams were mobilized.

Initial Response Phase August 30–31, 2005: Inadequate Situation Awareness

Governor Blanco stated that between 50,000 and 100,000 people have stayed behind in New Orleans, and she urged local officials to complete the evacuation of the city. However, due to the flooding most of the remaining residents were marooned. Evacuation plans based on school busses have failed to be implemented because the bus drivers were never trained to respond to their jobs during a disaster, and most of them have left the city. The public busses were used for evacuation initially, but most of the drivers refused to return to the city in the storm due to safety concerns, so the busses are at shelters in outlying areas. Many local residents with boats had begun a rescue operation in the flooded areas of the city. Coast Guard helicopters joined the search operations, which FEMA USARs deployed.

Mayor Ray Nagin and key staff members had taken shelter on the top floors of a hotel. They never opened the city's Emergency Operations Center, and had not taken the trained emergency management staff with them. The mayor's copy of the city's emergency plan was in the trunk of his car, underwater.

By midday on August 30, 2005, DHS Secretary Chertoff became aware that the levee system had failed and that there was no possibility of plugging the gap, raising the concern that essentially Lake Pontchartrain was going to start draining into the city. At the Pentagon, spokesman Lawrence Di Rita stated that the states affected by Hurricane Katrina had adequate National Guard units to handle the hurricane needs. Contradicting statements were made by Councilwoman Jackie Clarkson, who reported that the French Quarter has been attacked and few police were attempting to control looting when they should be used for search and rescue while there were still people on rooftops. The FEMA USAR teams were deploying from their air bases to Lafayette, Louisiana, from where they would rotate into the New Orleans area.

A complicating factor was the failure of the local first responder services. Fire personnel and equipment were marooned by the floodwaters. Almost 80% of the NOPD had been rendered homeless by the storm and subsequent flooding, and had gone to care for their families. Constitutional limitations on Federal response precluded the use of Federal troops without usurping Governor Blanco's control of the National Guard. The specter of a Republican president declaring a southern state with a Democratic governor in a state of insurrection deterred any action to send Federal troops (Lipton, Schmidtt, and Shanker, 2005). Meanwhile the USS *Bataan*, a 844 foot ship designed to dispatch Marines in amphibious assaults, with helicopters, doctors, hospital beds, food, and water rode out the storm and then followed it toward shore, awaiting relief orders. Helicopter pilots flying from its deck were some of the first to begin airlifting stranded New Orleans residents. But the *Bataan*'s hospital facilities, including six operating rooms and 600 beds for patients, remained unused, along with its complement of 1,210 sailors (Talking Points Memo, 2005).

DHS Secretary Chertoff declared Katrina an "Incident of National Significance, triggering for the first time a coordinated federal response to states and localities overwhelmed by disaster." The declaration was the first use of the new DHS National Response Plan (Talking Points Memo, 2005), and disclosed some of its weaknesses. This post 9/11 plan assumed that local first responders would initiate response and be supported by Federal assets. While this system worked well in Florida, Mississippi, and Alabama, the magnitude of the disaster in New Orleans overwhelmed Louisiana's capabilities.

By the evening of August 30, 2005, the first two USAR teams had arrived in Louisiana, and the convoy followed into Metairie by 7:45 a.m. on August 31. By noon the water rescue teams were in New Orleans at Causeway and Airline Boulevard to begin their search and rescue mission. On that day three California-based USARs rescued almost 300 people.

On August 31, 2005, conditions deteriorated in the Superdome in New Orleans, where 10,000 people had gone to seek refuge. The overcrowded and undersupplied structure was the site of frustration and anger. Governor Blanco ordered the evacuation of the people, and authorized commandeering busses to move them. The first people left that evening, bound for the Astrodome in Houston, Texas, 350 miles away. In the melee surrounding the exodus, at least three people died, including one man who jumped 17 m to his death. Since there was inadequate sanitation, the stench was described as overwhelming. At least 3,000 people went to the closed and locked Convention Center and broke in to seek shelter. As it had never been used as a shelter, the center had no supplies, and the water system was damaged by the storm. However, as it was in a dry area of the city, it was an attractive shelter.

Mayor Nagin ordered the 1,500 member NOPD force to stop its search and rescue duties and return to law enforcement. The mayor placed the city under curfew, and sent the police to quell the looting that was widespread. He requested additional federal assistance with search and rescue.

President Bush moved to Washington, DC to start meeting with a task force that would coordinate the work of 14 federal agencies involved in the relief effort. The Jefferson Parish emergency director, Walter Maestri, declared that the food and water supply had been used up, and that FEMA and other national agencies were not delivering the help nearly as fast as it was needed. Former Mayor Sidney Barthelemy estimated that 80,000 residents were trapped in the flooded city of New Orleans and urged President Bush to send more troops. At 9 p.m. FEMA Director Brown stated that Hurricane Katrina was much bigger than anyone expected.

Rescue Phase September 1, 2005: Disorganization

At 8 a.m., President Bush reiterated that nobody anticipated the breach of the levees. Although FEMA had been in New Orleans for the past 3 days, Terry Ebbert, New Orleans Homeland Security Director, complained that there had still been no command and control center established (CBC News, 2005) although the city had failed to open its own EOC facility. At 2 p.m., Mayor Nagin of New Orleans issued a "desperate SOS" to the federal government, since the city was out of resources, and the people who broke into the convention center needed more buses for evacuation purposes.

At that time New Orleans reportedly descended into anarchy. Looting and carjacking were widespread, resulting in the deployment of 30,000 additional National Guard personnel. Unknown people had looted guns from sporting goods stores and were firing on the rescuers who were in helicopters and boats. As a result, the USARs and Coast Guard were ordered back to camp to await reinforcements from law enforcement to protect them. Fires were breaking out as HAZMAT in the water contacted open flames. Corpses were lying out in the open, awaiting the rescue of the living before the collection of the dead.

On September 2, the governor declared a medical emergency, permitting out-of-state medical personnel to practice in the state. The federal Disaster Medical Assistance Teams (DMATs) began providing medical care at the Superdome and the airport, from where the most seriously ill were airlifted to other communities. Claims of lawlessness were disputed by FEMA Director Michael Brown, who learned of evacuees at the convention center only this day, at which time he directed all available resources to get to the convention center. With the arrival of the mutual aid law enforcement personnel, the USARs were back in the community, providing water to stranded residents, and rescuing over 500 more people by September 3. By September 4, most residents had been found, and rescues continued, with three people and five dogs taken to shelters. Helicopter-based rescues began, with the USARs rescuing 33 more people by the end of the week (Schapelhouman, 2005).

By September 2, the Governor had declined the federal offer to take over the management of the disaster response, leaving the responsibility in the hands of the city and the state. Offers of mutual aid through the Emergency Management Assistance Compact were vetted by the Louisiana officials, and selections were made based on lowest cost and proximity to the disaster. Congress approved and President Bush signed an initial $10.5 billion aid package for immediate rescue and relief efforts. However, the Congressional Black Caucus, along with the NAACP, Black Leadership Forum, and the National Urban League, expressed dismay over the sluggish relief efforts in New Orleans, citing the poverty of the victims as a primary reason for the delay.

On Friday, September 9, former Secretary of State Colin Powell, in a 20/20 interview, criticized the response at all levels of the government to Hurricane Katrina, saying "When you look at those who weren't able to get out, it should have been a blinding flash of the obvious to everybody that when you order a mandatory evacuation, you can't expect everybody to evacuate on their own. These are people who don't have credit cards; only one in 10 families at that economic level in New Orleans have a car. So it wasn't a racial thing, but poverty disproportionately affects African-Americans in this country. And it happened because they were poor" (Talking Points Memo, 2005).

ACRONYMS

1GW	First-generation warfare
2GW	Second-generation warfare
2PAM	Atropine auto-injector
3GW	Third-generation warfare
4GW	Fourth-generation warfare
9/11	September 11, 2001
9-1-1	US emergency services phone number
AEDPA	Anti-Terrorism and Effective Death Penalty Act
ALOHA	Aerial Locations of Hazardous Atmospheres
ANFO	Ammonium nitrate and fuel oil
ARGUS rapid alert system	European Union Rapid Alert System
ATLANTIC BLUE	A major international counterterrorism exercise involving the UK, USA, and Canada
ATSSA	Air Transportation and System Stabilization Act
AWAC	Airborne Warning and Control System
BATWING	Bay Area Terrorism Working Group
BAYMACS	Bay Area Mutual Aid Communications System
BBC	British Broadcasting Company
$C4^I$	Command, control, communications, computing, and information systems ($C4^I$)
CAD	Computer-aided dispatching
CADRE	Collaborative Agencies Disaster Relief Effort
CATS	Catastrophe Assessment Tool Set
CAM	Chemical agent monitors
CAMEO	Computer-aided management of emergency operations
CBC	Canadian Broadcasting Company
CBRNE	Chemical, biological, radiological, nuclear and explosive
CCTV	Closed-circuit television
CEP	Civil emergency planning
CERT	Community Emergency Response Team
COBRA	Cabinet Office Briefing Rooms (UK)
CNEN	National Nuclear Energy Commission (Brazil)
CNN	Cable News Network
DCDEP	Directorate for Civil Defense and Emergency Planning
DEMA	Danish Emergency Management Agency
DHS	Department of Homeland Security
Dirty Bomb	Chemical or radiological dispersal device
DNA	Deoxyribonucleic acid
DMAT	Disaster Medical Assistance Team

DOE	Department of Energy
EADRCC	Euro-Atlantic Disaster Response Coordination Centre
EAPC	Euro-Atlantic Partnership Council
EAS	Emergency Alert System
EBS	Emergency Broadcast System
EMERCOM	Ministry of Russian Federation for Civil Defense, Emergencies, and Elimination of Consequences of Natural Disasters
EOC	Emergency Operations Center
EPA	Environmental Protection Agency
ESF	Emergency Support Functions
EU	European Union
EUCENTRE	European Centre for Training and Research in Earthquake Engineering
EURO 2004	European Football (Soccer) Championship, held in Portugal
FBI	Federal Bureau of Investigation
FDNY	Fire Department of New York
FEMA	Federal Emergency Management Agency
FI	Field information
FIFA	Fédération Internationale de Football Association
FIRESCOPE	Firefighting Resources of California Organized for Potential Emergencies
FTIR	Fourier Transform Infra-Red spectroscopy
G8	Group of 8—International forum for the governments of Canada, France, Germany, Italy, Japan, Russia, the UK, and the USA
GIS	Geographic Information Systems
Gy	Gray (unit) for absorbed dose of radiation
GW	Generational Warfare
HAZMAT	Hazardous materials
HEPA	High-efficiency particulate air (filter)
HEU	Highly enriched uranium
HMTD	Hexamethylene Triperoxide Diamine
HPAC	Hazard Prediction and Assessment Capability
IAEA	International Atomic Energy Agency
ICS	Incident Command System
ICT	Information Communication and Technology (The Netherlands)
IDF	Israeli Defense Force
IED	Improvised Explosive Device

IND	Improvised nuclear device
Inter alia	A legal term meaning "among other things"
IRA	Irish Republican Army
IS	International Staff
JRIES	Joint Regional Information Exchange System
LED	Light-emitting diode
LEON Principle	Lowest, effective level of care principle (Norway)
LOCC	Country Operational Coordination Centre (The Netherlands)
MACS	Mutual Aid Coordination System
MIC	Monitoring Information Center
MMRS	Metropolitan Medical Response System
MMST	Metropolitan Medical Strike Team
MOPP gear	Mission-oriented protective posture gear
NATO	North Atlantic Treaty Organization
NATO-EU CAP	Joint NATO–EU Capabilities Group
NATO-EU PRO	Coordinated NATO–EU Training Program
NATO-EU RES	Coordinated NATO–EU Research Program
NATO-EU SEC	Joint NATO–EU Homeland Security Clearing House (NATO-EU SEC
NBC-CREST	Nuclear Biologic Chemical Casualty and Resource Estimation Support Tool
NBC weapons	Nuclear, biological, and chemical weapons
NCC	Network Control Center
NCC	National Control Center (The Netherlands)
NGO	Nongovernmental organization
NIMS	National Incident Management System
NIST	National Institute of Standards and Technology
NOAA	National Oceanic and Atmospheric Administration
NOPD	New Orleans Police Department
NRF	NATO Response Force
NRP	National Response Plan
NYPD	New York Police Department
OCHA	Office for the Coordination of Humanitarian Affairs
OEM	Office of Emergency Management
OODA loop theory	Colonel John Boyd's Observation, Orientation, Decision, and Action loop theory, also known as Boyd cycles
OPCW	Organization for the Prohibition of Chemical Weapons
OPLAN	Operation plan
OPLAN ABLE GUARDIAN	NATO plan for fight against terrorism
OREMS	Oak Ridge Evacuation Modeling System

OSST	Office for Security Science and Technology
PANY	Port Authority of New York
PAPD	Port Authority Police Department
PCR	Polymerase chain reaction
PDRM 82	Portable Radiological Dose Rate Meter
PET	Security Intelligence Service (Denmark)
PIO	Public Information Officer
PPE	Personal protective equipment
RCC	Rescue Coordination Centers
RSC	Rescue subcenter
R&D	Research and Development
RDD	Radiological dispersal device
SAFECOM	Communications program of the Department of Homeland Security's Office for Interoperability and Compatibility (OIC)
SAME	Specific Alert Message Encoding
SAR	Search and Rescue Service (Norway)
SEMS	Standardized Emergency Management System
SHAPE	Supreme Headquarters, Allied Powers Europe
START	Simple Triage and Rapid Treatment
S&T	Science and technology
STS-CNAD	Science, Technology, and Society Conference on National Armaments Directors
SWAT	Special Weapons and Tactics Team
TATP	Triacetone Triperoxide
Tbq	Terabecquerel: a unit of radioactivity
TEWG	Terrorism Early Warning Group
TIC	Toxic industrial chemicals
TOPOFF	Top Officials (a terrorism response exercise conducted in the USA)
UEFA	Union of European Football Associations
UN	United Nations
USA PATRIOT Act	Uniting and Strengthening America by Providing Appropriate Tools Required to Intercept and Obstruct Terrorism Act of 2001
USAR	Urban Search and Rescue
UNHCR	UN High Commission on Refugees
UNOCHA	UN Office for the Coordination of Humanitarian Affairs
VBIED	Vehicle-borne improvised explosive device
VNC	Voluntary National Contribution
VOAD	Voluntary Organizations Active in Disasters
WiFi	Wireless fidelity

WFP	World Food Program
WMD	Weapons of mass destruction
WMDi	Weapons of mass disruption
WMK	Weapons of mass killing
WMV	Weapons of mass victimization

REFERENCES

Chapter 1

BBC, 2006, Iraq makes terror 'more likely,' Feb 28. http://news.bbc.co.uk/1/hi/world/middle_east/4755706.stm.

Dalgaard-Nielsen, A. and Hamilton, D.S. eds., 2006, *Transatlantic Homeland Security*. Routledge, Oxford.

Chapter 2

2.0–2.1

CBC News, 2005, New Orleans official criticizes FEMA, Sept 1. http://www.cbc.ca/world/story/2005/09/01/Ebbert_FEMA20050901.html.

Dhara, V.R. and Dhara, R., 2002, The Union Carbide disaster in Bhopal: a review of health effects. *Archives of Environmental Health*, 57, 391–404.

Governor's Office, 2005, Governor Blanco declares state of emergency, Aug 26. http://www.gov.state.la.us/index.cfm?md=newsroom&tmp=detail&catID=1&articleID=776&navID=3.

Honore, Lt. Gen. Russel, 2005, Commander of the Katrina Joint Task Force, Department of Defense, Press conference, Sept. 1.

Kean, T.H. and Hamilton, L.H., 2004, *The 9/11 Report, The National Commission on Terrorist Attacks Upon the United States*. St. Martin's Press, New York.

Steinhäusler, F., 2006, Risk-based emergency preparedness against terror attacks with WMD, Argumente u. Materialien zum Zeitgeschehen (Publ.: Akademie f. Politik u. Zeitgeschehen, Munich, Germany; eds. Volker Foertsch, Klaus Lange), No. 50, pp. 79–88.

Steinhäusler, F., 2005, Chernobyl and Goiania lessons for responding to radiological terrorism. *Health Physics*, 89(5), 566–574.

Steinhäusler, F. and Edwards, F., eds., 2005, *NATO and Terrorism: Catastrophic Terrorism and First Responders—Threats and Mitigation*. Springer, Dordrecht, The Netherlands.

Wyden, P., 1984, *Day One Before Hiroshima and After*. Simon & Schuster, New York.

2.2

CNN, 2003, 20 years later, Lebanon bombing haunts, http://www.cnn.com/2003/WORLD/meast/10/21/lebanon.anniv.ap/.

CNN Interactive, 1995, Oklahoma City Bombing Trials, http://www.cnn.com/US/9703/okc.trial/timeline/1995.html.

Parker, J., 2006, Final Rajneeshee Sentences in Murder Plot, kgw.com, http://www.kgw.com/news-local/stories/kgw_013005_news_Rajneesh_follower_sentencing.54f8650d.html.

Raymond, N.J., 2006, Iraq: PHR documentation of chemical weapons against Kurds by Hussein regime's Anfal Campaign, Physicians for Human Rights, http://physiciansforhumanrights. org/library/news-2006-08-24a.html.

Staten, C.L., 1999, EmergencyNet Exclusive: questions and answers on bio-warfare/bio-terrorism (Q&A) with Dr. Ken Alibek, http://www.emergency.com/1999/alibek99.htm.

2.3

Department of Homeland Security, 2004, Homeland Security Grant Guidance, p. 38. Accessed through http://www.hsem.state.mn.us/grants/04_hsgp/letpp_grant.pdf.

Homeland Security Advisory Council, 2004, Intelligence and Information Sharing Initiative, "Final Report," December.

LLIS Intelligence and Information Sharing Initiative: Homeland Security Intelligence Requirements Process, 2005, p. 2. http://www.dhs.gov/xlibrary/assets/Final_LLIS_Intel_ Reqs_Report_Dec05.pdf.

Maimon A., 2006, Federal antiterror funds go toward sharing information. *Las Vegas Review-Journal*, August 4, 5B.

Steinhäusler, F. and Edwards, F., eds., 2005, *NATO and Terrorism: Catastrophic Terrorism and First Responders—Threats and Mitigation.* Springer, Dordrecht, The Netherlands.

2.4

Antiterrorism and Effective Death Penalty Act of 1996, 1996, Pub. L. No. 104–132, 110 Stat. 1214.

Ballard, J.D., 2005, *Terrorism, Media, and Public Policy: The Oklahoma City Bombing.* Hampton Press, Cresskill, NJ.

Ballard, J.D. and Mullendore, K.B., 2004, The presumption of response to terrorism: potential legal aspects relative to first responders. Proceedings of NATO Advanced Research Workshop on Catastrophic Terrorism and First Responders: Threats and Mitigation in Neuhaussen-Stuttgart, Univeristat Salzburg, Austria.

Barkett, J.M., 2003, If terror reigns, will torts follow? *Widener Law Symposium*, 9, 485–543.

Berkovitz, D.M., 1989, Price-Anderson Act: model compensation legislation?—the sixty-three million dollar question. *Harvard Environmental Law Review*, XIII.

Berman, H.J. and Reid, C.J., 1996, The transformation of English science: from Hale to Blackstone. *Emory Law Journal*, 45, 437–522.

Bhoumik, A., 2005, Democratic responses to terrorism: a comparative study of the United States, Israel, and India. *Denver Journal of International Law and Policy*, 33, 285–345.

Bina, M.W., 2005, Private military contractor liability and accountability after Abu Ghraib. *John Marshall Law Review*, 38, 1237–1262.

Culhane, J.G., 2003, Tort, compensation, and two kinds of justice. *Rutgers Law Review*, 55, 1027–1107.

Floering, W., 2002, The September 11th Victim Compensation Fund of 2001: a better alternative to litigation? *Journal of the National Association of Administrative Law Judges*, 22, 195–221.

Goldman, J.J., 2003, Ruling opens way for 9/11 survivors and victims to sue, Truthout, December. http://www.truhout.org/docd_03/printer_091103F.shtml.

Guiora, A.N., 2005, Legislative and policy responses to terrorism: a global perspective. *San Diego International Law Journal*, 7, 125–172.

Hamblett, M., 2002, Victim's lawsuits flood court to meet one-year time limit. *New York Law Journal.* Lexis Academic Universe.

Kay, J., 2001, Moratorium declared on terrorism suits. *Fulton County Daily Report*. Lexis Academic Universe.

Kellman, B., 1999, Catastrophic terrorism: thinking fearfully, acting legally. *Michigan Journal of International Law*, 20, 537–564.

Kingshott, B., 2003, Terrorism: the new religious war. *Criminal Justice Studies*, 16(1), 15–27.

Manns, J., 2003, Insuring against terror? *Yale Law Journal*, 112, 2509–2551.

Maogoto, J.N., 2005, Countering terrorism: from wigged judges to helmeted soldiers— legal perspectives in America's counterterrorism responses. *San Diego International Law Journal*, 6, 243–294.

Mostaghel, D.M., 2001, Wrong place at the wrong time, unfair treatment? Aid to victims of terrorist attacks. *Brandeis Law Journal*, 40, 83–120.

Mullenix, L.S., 2004, The future of tort reform: possible lessons from the World Trade Center Victim Compensation Fund. *Emory Law Journal*, 53, 1315–1347.

Mullenix, L.S. and Stewart, K.B., 2002, The September 11th Victim Compensation Fund: fund approaches to resolving Mass Tort Litigation. *Connecticut Insurance Law Journal*, 9, 121–152.

Neumeister, L., 2003, Judge backs 9/11 lawsuits. Associated Press/*Baltimore Sun*. http://www.baltimoresun.com/business/bal-lawsuits0909,0,5214287,print.story?coll=bal-business-headlines.

Ramirez, J., 2002, The Victims Compensation Fund: a model for future mass casualty situations. *Transportation Law Journal*, 29, 283–298.

Reynolds, M.L., 1996, Landowner liability for terrorist acts. *Case Western Law Review*, 47, 155–206.

September 11th Compensation Fund, 2002, 49 USC 44903.

Steinhäusler, F. and Edwards, F., eds., 2005, *NATO and Terrorism: Catastrophic Terrorism and First Responders—Threats and Mitigation*. Springer, Dordrecht, The Netherlands.

Terrorism Risk Insurance Act of 2002, 2003, 15 USCA 248, 16109, 6701.

Uniting and Strengthening America by Providing Appropriate Tools to Intercept and Obstruct Terrorism (USA PATRIOT ACT) of 2001, Pub. L. No. 107-56, 115 Stat. 272, 2001.

Wattellier, J.J., 2004, Comparative legal responses to terrorism: lessons from Europe. *Hastings International and Comparative Law Review*, 27, 397–419.

Weinzierl, J., 2004, Terrorism: its origin and history. In Nyatepe-Coo, A.A. and Zeisler-Vralsted, D. eds., *Understanding Terrorism: Threats in an Uncertain World*. Pearson Prentice Hall, Upper Saddle River, NJ.

Chapter 3

3.0

Steinhäusler, F. and Edwards, F., eds., 2005, *NATO and Terrorism: Catastrophic Terrorism and First Responders—Threats and Mitigation*. Springer, Dordrecht, The Netherlands.

3.1

Associated Press, 2004, National alert system in disarray, June 28. http://www.cbsnews.com/stories/2004/06/28/terror/printable626452.shtml.

Bush, G.W., 2002, Homeland Security Presidential Directive-3. The White House, Washington, D.C. http://www.whitehouse.gov/news/releases/2002/03/20020312-5.html.

Department of Homeland Security, Community Emergency Response Team, https://www.citizencorps.gov/cert/.

Ding A., 2006, A theoretical model of public response to the homeland security advisory system, *JDMS*, 3(1), 45–55.
http://www.scs.org/pubs/jdms/vol3num1/JDMSvol3no1Ding45-55.pdf.

Federal Communications Commission, no date, Fact sheet: emergency broadcast system. http://www.fcc.gov/eb/easfact.html.

Federation of American Scientists, 2007, Emergency Broadcast System. http://www.fas.org/ nuke/guide/usa/c3i/ebs.htm.

Lopes, R., 2002, Eleven 'C's' of community disaster education, American Red Cross, Washington, D.C.

NOAA, 2007, NWR receiver consumer information. http://www.nws.noaa.gov/nwr/ nwrrcvr.htm.

NOAA, 2005, Frequently asked questions about NOAA weather radio. http://www.srh. noaa.gov/bro/nwrfaq.htm.

Partnership for Public Warning, The Common Alerting Protocol: an open standard for interoperability in all-hazard warning. http://www.ppw.us/ppw/cap.html.

Rein, L., 2006, All ears for a blast from the past. *Washington Post*, January 25, p. A01.

72hours.org., Are You Prepared?, San Francisco Office of Emergency Services and Homeland Security, http://www.72hours.org/sirens.html.

Simon, E., 2004, Expert panel: national alert system in disarray, Associated Press, June 27. http://www.signonsandiego.com/news/nation/20040627-0939-emergencywarning.html.

Vedantam, S., 2006, Repeated warnings have diminishing returns. *Washington Post*, November 20, p. A02.

3.2

Associated Press, 2005, New Orleans flees as Katrina approaches Gulf Coast, August 28. http://www.usatoday.com/weather/stormcenter/2005-08-28-katrina-gulf_x.htm.

California Governor's Office of Emergency Services. Standardized Emergency Management System. http://www.oes.ca.gov/Operational/OESHome.nsf/Content/B494353521089544 88256C2A0071E038?OpenDocument.

CNN, 2003, Ridge tries to calm America's nerves, February 14. http://www.cnn.com/ 2003/ALLPOLITICS/02/14/homeland.security.

Department of Homeland Security, 2004, National Response Plan. http://www.dhs.gov/ xlibrary/assets/NRP_FullText.pdf.

Federal Emergency Management Agency, 2006a, Robert T. Stafford Disaster Relief and Emergency Assistance Act, as amended. http://www.fema.gov/about/stafact.shtm.

Federal Emergency Management Agency, 2006b, Urban Search and Rescue Teams. http://www.fema.gov/emergency/usr/.

Federal Emergency Management Agency, no date, Incident Command System. http://training.fema.gov/EMIWeb/IS/is100.asp.

Federal Emergency Management Agency, 2004. Hurricane Pam Exercise Concludes. http://www.fema.gov/news/newsrelease.fema?id=13051.

Hsu, S.S., 2006, Katrina Report Spreads Blame. *Washington Post*, February 12, p. A01.

National Voluntary Organizations Active in Disaster, http://www.nvoad.org/.

State of California, Emergency Medical Services Authority, Disaster Medical Assistance Teams. http://www.emsa.ca.gov/Dms2/dmatinfo.asp.

The City of Oklahoma City, 1996, *Alfred P. Murrah Federal Building Bombing, April 19, 1995*. Fire Protection Publications, Stillwater, OK, p. 34.

USA Today, 2005, Evacuation worked, but created a highway horror, September 25. http://www.usatoday.com/news/nation/2005-09-25-evacuation-cover_x.htm.

Vartabedian R., 2006, Officials add $6 billion to price tag of levees. *Los Angeles Times*, March 31, p. A01.

Volunteer Center of Silicon Valley, CADRE. http://www.vcsv.us/cadre.shtml.

Winslow, F.E., 2001, Planning for weapons of mass destruction/nuclear, biological and chemical agents: a local/federal partnership. *In Handbook of Crisis and Emergency Management*, Ali Farazmand, ed. Marcel Dekker, New York.

3.3

Brower Window Films, 2007, Safety/Security. http://www.tintandgraphics.com/faqs/faqs.html.

Google Earth: Explore, Search and Discover, 2007. www.earth.google.com.

Kennedy, G. and Klein, L., 2002, Why the Towers Fell. WGBH, Boston (videotape).

World Trade Center Building Code Task Force. http://www.nyc.gov/html/dob/html/guides/wtc.shtml.

3.4

BAYMACS, 2003, Introducing BAYMACS. http://www.oes.ca.gov/Operational/OESHome.nsf/PDF/psrspc_SCC/$file/SCC_07-31-03.ppt.

Bush, G.W., 2003, Homeland Security Presidential Directive-8: National Preparedness. http://www.whitehouse.gov/news/releases/2003/12/20031217-6.html.

ConsumerAffairs.com, 2005, New Orleans phones, cell networks still down, September 2. http://www.consumeraffairs.com/news04/2005/katrina_phones.html.

Dawes, S., Birkland T., Tayi, G.K., and Schneider, C.A., 2004, *Information, Technology, and Coordination: Lessons from the World Trade Center Response*. Center for Technology in Government, Albany, NY.

Department of Homeland Security, 2004, SAFECOM: Interoperability continuum. http://www.safecomprogram.gov/NR/rdonlyres/65AA8ACF-5DE6-428B-BBD2-7EA4BF44FE3A/0/Continuum080106JR.pdf.

Edwards, F. and Goodrich, D., 2006, *Post Drill Meeting Report: Caltrans Contraflow Exercise*. Mineta Transportation Institute, San José, CA, November 19.

Federal Emergency Management Agency, 2005, NIMS Integration Center, Resource Typing, July. http://www.fema.gov/pdf/emergency/nims/resource_typing_qadoc.pdf.

FIRESCOPE, (no date), http://www.firescope.org/.

Incident Command System, 2004, Field Operations Guide, ICS 420-1, FIRESCOPE, June.

Interpol, 2006, Public Safety and Terrorism, August 22. http://www.interpol.int/Public/Terrorism/default.asp.

National Commission on Terrorist Attacks Upon the United States, *9/11 Commission Report*, Chapter 9. http://www.9-11commission.gov/report/911Report_Ch9.htm.

NIST, 2005, NIST NCSTAR 1: Federal Building and Fire Safety Investigation of the World Trade Center Disaster. Final Report on the Collapse of the World Trade Center Towers, Department of Commerce, Washington, DC. http://wtc.nist.gov/NISTNCSTAR1Collapse ofTowers.pdf.

Nurbaiti, A., 2005, Indonesia: "I'm Alive" Project continues to reunite families in Aceh. *The Jakarta Post*, April 20. http://www.asiamedia.ucla.edu/tsunami/article.asp?parentid=23261.

O'Hare, Commander F.M., 2002, New York City Police Department Transit Bureau. PowerPoint presentation, August 25.

Reyes, E., 2006, Basic Public Safety Communications and Interoperability, PoliceOne.com. http://www.policeone.com/police-technology/mobile-data/articles/127284/.

Steinhäusler, F. and Edwards, F., eds., 2005, *NATO and Terrorism: Catastrophic Terrorism and First Responders—Threats and Mitigation.* Springer, Dordrecht, The Netherlands.

Swider, P., 2003, Patrolling with WiFi, WiFi Planet Insights, November 10. http://www.wi-fiplanet.com/columns/article.php/3106771.

Tropos, Public Safety. http://www.tropos.com/public_safety.html.

Wills, T., 2007, Anaheim's Breadcrumbs, Anaheim Fire Department presentation, Region IX Metropolitan Medical Response System quarterly meeting, January 24.

Chapter 4

4.0–4.1

BBC News, 2005, Bomb Suspect Arrested by Police, July 27. http://news.bbc.co.uk/2/hi/uk_news/4720027.stm

Bretschneider, 2006, NATO Euro-Atlantic Disaster Response Coordination Center. Presentation at the NATO STS/ND workshop, Ericera, Portugal, March.

Brown, G., The Rt. Hon. MP, 2006, Securing Our Future. Lecture, Royal United Services Institute for Defence Studies (RUSI), London.

Leppard, D. et al., 2005, Shattered. *Sunday Times*, July 10, pp. 13–14.

Metropolitan Police Service, 2005, Press Conference, July 26. http://cms.met.police.uk.

Scanlon, J., 2006, Dealing with the Dead. Presentation at the International Association of Emergency Managers Annual Conference, Orlando, Florida, November.

Sugar, R.D., 2006, Trends and Opportunities U.S./U.K. Defense Co-Operation in a New Environment. Lecture, Royal United Services for Defence Studies (RUSI), London.

Tendler, S. and O'Neill, S., 2005, 6,000 officers deployed to beat Thursday attack. *The Times*, July 29, p. 6.

UK Government, 2004, Civil Contingencies Act. http://www.opsi.gov.uk/acts/acts2004/20040036.html.

US Department of Homeland Security, 2005a, Fact Sheet: TOPOFF 3 Exercising National Preparedness. www.dhs.gov/dhspublic/display?content=4410.

US Department of Homeland Security, 2005b, Office of Inspector General. A Review of the Top Officials, 3 Exercise, OIG-06-07, p. 13–14.

Woods, R. et al.,, 2005, Wanted: dead or alive. *The Sunday Times*, July 24, p. 15.

4.3

Werner, H., 2006, National Emergency Preparedness in Germany. NATO STS-CNAD Workshop, Ericera, Portugal, March 3.

4.4

Contact/Working Group on Emergency Preparedness to Deal with Means of Mass Destruction, 2002, Proposal for an overall national plan for preparedness against means of mass destruction in Norway. Drawn up by the appointed, according to terms of reference laid down by the Ministry of Justice on 7.

Forsvarsdepartementet, 2004, Further modernization of the Norwegian armed forces 2005–2008, March 12. http://odin.dep.no/fd/english/news/news/010001-070030/dok-bn.html.

NRPA, 2006, Ny kgl.res: Atomberedskap—sentral og regional organisering, NRPA.NO, 22.02.

4.5

Dalgaard-Nielsen, A. and Hamilton, D.S. eds., 2006, *Transatlantic Homeland Security*. Routledge, Oxford, UK.

Hedge, A. 1987, Major hazards and behaviour, in Singleton, W.T. and Hovden, J., eds., *Risk and Decisions*. Wiley, Chichester, UK, pp. 148–149.

Strat, 2005, Concepts. 24. http://odin.dep.no/filarkiv/254514/Relevant%20force.pdf.

U.S. Office of Homeland Security, 2002, National Strategy for Homeland Security, Washington, DC. www.whitehouse.gov/homeland/book/nat_strat_hls.pdf.

4.6

Campeao, A., 2006, Emergency Preparedness in Portugal and Spain. Presentation at STS-CNAD, Ericiera, Portugal.

4.7

California Governor's Office of Emergency Services, 2006, Standardized Emergency Management System. http://www.oes.ca.gov/Operational/OESHome.nsf/Content/B494 35352108954488256C2A0071E038?OpenDocument.

California Fire and Rescue Training Authority, 2000, California Rescue Systems 1 Manual. California Fire and Rescue Training Authority, McClellan, CA.

California Urban Search and Rescue Task Force 3, 2005, *Hurricane Katrina After-Action*. Report, USAR CA-3, Menlo Park, CA.

Department of Homeland Security, 2006, National Response Plan. http://www.dhs.gov/xprepresp/committees/editorial_0566.shtm.

FIRESCOPE, no date, California Incident Command System, (ICS). http://www.firescope.org/.

Schapelhouman, H., Menlo Park Fire District Chief, 2006, Presentation at the STS-CNAD Workshop in Ericiera, Portugal.

4.8

Thomas, T., 1995, EMERCOM: Russia's emergency response team. *Low Intensity Conflict and Law Enforcement*, 4(Autumn), 227–236. http://www.fas.org/nukc/guidc/russia/agency/rusert.htm.

4.9

Sagiv, Y., 2006, Presentation at STS-CNAD workshop, Ericiera, Portugal.

Stanford, 2006, Best anthrax treatment: rapid diagnosis and antibiotics. Stanford Report. http://news-service.stanford.edu/news/2006/february22/med-anthrax-022206.html.

4.10

Bretschneider, G., 2006, Presentation at STS-CNAD workshop, Ericeira, Portugal, March 3.

Butterfield, A., 2006, Presentation at STS-CNAD workshop, Ericeira, Portugal.

4.11

European Union, 2007a, Common emergency communication and information system. *Civil Protection*, Jan 3. http://ec.europa.eu/environment/civil/cecis.htm.

European Union, 2007b, Community Civil Protection Action Programme. *Civil Protection*, Jan 3. http://ec.europa.eu/environment/civil/prote/cp14_en.htm.

European Union, 2007c, Community cooperation in the field of civil protection. *Civil Protection*, Jan 3. http://ec.europa.eu/environment/civil/prote/cp01_en.htm.

European Union, 2007d, Cross border pilot projects. *Civil Protection*, Jan 3. http://ec.europa.eu/environment/civil/prote/crossborder.htm.

12

Trapp R., 2006, Presentation, STS-CNAD workshop, Ericeira, Portugal.

13

Guiora, A., 2007, Cost-Benefit of Counterterrorism. Lecture at San José State University, February 15.

Herbert, A.J., 2003, The Nato response force. *Air Force Magazine* Online, April, 86(4). http://www.afa.org/magazine/april2003/0403NATO.asp.

Olsen, K., 1999, Aum Shinrikyo: once and future threat. *Emerging Infectious Diseases*, 5(4), 513–516.

Parsons, C., 2007, Chlorine bombs mark new guerrilla tactics: US, Reuters, Feb 22. http://www.reuters.com/article/topNews/idUSKRA14854020070222.

Chapter 5

1

Carafano, J.J., Czerwinski, J.J., and Weitz, R., 2006, Homeland security technology, global partnerships, and winning the long war: backgrounder #1977. The Heritage Foundation, October 5. http://www.heritage.org/Research/HomelandDefense/bg1977.cfm.

4

Bush, G.W., 2003, Homeland Security Presidential Directive-7: Critical Infrastructure Identification, Prioritization, and Protection, December 17. http://www.whitehouse.gov/news/releases/2003/12/20031217-5.html (accessed 1/29/07).

Department of Homeland Security, 2003, The National Strategy for the Physical Protection of Critical Infrastructures and Key Assets, February. http://www.dhs.gov/xlibrary/assets/Physical_Strategy.pdf.

Department of Homeland Security, 2006, National Infrastructure Protection Plan. http://www.dhs.gov/xlibrary/assets/NIPP_Plan.pdf.

Palm, 2005, You can learn about Treo Smartphones. http://www.palm.com/intl/landing/treo_intl.epl.

Chapter 6

6.2

Edwards, F.L., 2006, Law enforcement response to biological terrorism: lessons learned from New Orleans after Hurricane Katrina. *Law Enforcement Forum Journal*, 6(2).

Steinhäusler, F. and Edwards, F., eds., 2005, *NATO and Terrorism: Catastrophic Terrorism and First Responders—Threats and Mitigation.* Springer, Dordrecht, The Netherlands.

Appendix 2

Antal, J. and Gericke, B., 2003, *City Fights: Selected Histories of Urban Combat from World War II to Vietnam*. Ballantine Books, New York.

Adams, R., McTernan, T., and Remsberg, C., 1990, *Street Survival: Tactics for Armed Encounters*. Caliber Press, Northbrook, IL.

Biddle, S., 2004, *Military Power: Explaining Victory and Defeat in Modern Battle*. Princeton University Press, Princeton, NJ.

Coram, R., 2002, *Boyd: The Fighter Pilot Who Changed The Art Of War*. Back Bay Books, New York.

Poole, J., 1999, *One More Bridge to Cross: Lowering the Cost Of War*. Posterity Press, Emerald Isle, NC.

Galula, D., 1964, *Counterinsurgency Warfare: Theory and Practice*. Frederick A. Praeger, Westport, CT.

Grossman, D., 1995, *On Killing: The Psychological Cost of Learning to Kill in War and Society*. Back Bay Books, New York.

Gudmundsson, B., 2004, *On Armor*. Praeger Publishers, Westport, CT.

Gudmundsson, B., 1989, *Stormtroop Tactics: Innovation In The German Army, 1914–1918*. Praeger Publishers, Westport, CT.

Hammel, E., 1992, *Six Days In June: How Israel Won The 1967 Arab-Israeli War*. ibooks, New York.

Hammes, T., 2004, *The Sling and the Stone: On War in the 21st Century*. Zenith Press, St. Paul, MN.

Hammond, G., 2001, *The Mind Of War: John Boyd and American Security*. Smithsonian Books, Washington, DC.

Larsen, E.V. and Peters, J.E., 2001, Preparing the US Army for Homeland Security: Concepts, Issues and Option. MR-1251-A, Rand Corporation. http://www.rand.org/pubs/monograph_reports/MR1251/index.html.

Lind, W., 1985, *Maneuver Warfare Handbook*. Westview Press, Boulder, CO.

Lind, W., Nightengale, K., Schmitt, J., Sutton, J., and Wilson, G., 1989, The changing face of war: into the fourth generation. *Marine Corps Gazette*, October.

Murray, K., 2004, *Training At The Speed Of Life*, Volume I. Armiger Publications, Gotham, NY.

Nagal, J., 2005, *Learning To Eat Soup With A Knife: Counterinsurgency Lessons from Malaya and Vietnam*. University of Chicago Press, Chicago, IL.

Nunn-Lugar-Domenici Domestic Preparedness Program, 1996, Defense Against Weapons of Mass Destruction Act of 1996 (PL 104–201).

Remsberg, C., 1990, *The Tactical Edge: Surviving High-Risk Patrol*. Caliber Press, Northbrook, IL.

Schneider, W., 2005, *Panzer Tactics: German Small-Unit Armor Tactics in World War II*. Stackpole Books, Mechanicsburg, PA.

Stubblefield, G. and Monday, M., 1994, *Killing Zone: A Professional's Guide to Preparing or Preventing Ambushes*. Paladin Press, Boulder, CO.

Van Creveld, M., 1991, *The Transformation Of War*. Free Press, New York.

Appendix 3

International Atomic Energy Agency, 1998, Proceedings International Conference Goiania—10 Years Later. Department of Nuclear Safety, Doc. No. IAEA-GOCP, A-1400 Vienna, Austria.

Appendix 4

Kean, T.H. and Hamilton, L.H., 2004, *The 9/11 Report, The National Commission on Terrorist Attacks Upon the United States*. St. Martin's Press, New York.

Appendix 5

CBC News, 2005, New Orleans official criticizes FEMA, Sept 1. http://www.cbc.ca/world/story/2005/09/01/Ebbert_FEMA20050901.html.

Fechter, M., 2004, Thanks to Hurricane Andrew, response swift. *Tampa Tribune*, August 22. http://www.tampatrib.com/nationworldnews/MGBIV6QN6YD.html.

Lipton, E., Schmidtt, E., and Shanker, T., 2005, Political issues snarled plan for troop aid. *New York Times*, September 9. http://www.nytimes.com/2005/09/09/national/nationalspecial/09military.html.

Schapelhouman, Chief Harold, 2005, CA-3 USAR Task Force, PowerPoint presentation at Business Continuity Planning: Hurricane Katrina Conference. NASA-Ames Research Center, Mountain View, CA, November 2.

Talking Points Memo, 2005, Hurricane Katrina Timeline, September 20. http://www.talkingpointsmemo.com/katrina-timeline.php.

INDEX